APOLLO 11

THE MOON LANDING IN REAL TIME

APOLLO 11

THE MOON LANDING IN REAL TIME

WITHDRAWN

IAN PASSINGHAM

PEN & SWORD
HISTORY

AN IMPRINT OF PEN & SWORD BOOKS LTD.
YORKSHIRE ~ PHILADELPHIA

First published in Great Britain in 2019 by
PEN AND SWORD HISTORY
An imprint of
Pen & Sword Books Ltd
Yorkshire – Philadelphia

Hardback ISBN 978 1 52674 856 0
Paperback ISBN 978 1 52675 761 6

Typeset in Times New Roman 11.5/14 by
Aura Technology and Software Services, India
Printed and bound in the UK by TJ International Ltd.

Pen & Sword Books Limited incorporates the imprints of Atlas, Archaeology,
Aviation, Discovery, Family History, Fiction, History, Maritime, Military, Military
Classics, Politics, Select, Transport, True Crime, Air World, Frontline Publishing,
Leo Cooper, Remember When, Seaforth Publishing, The Praetorian Press,
Wharncliffe Local History, Wharncliffe Transport, Wharncliffe True Crime and
White Owl.

For a complete list of Pen & Sword titles please contact
PEN & SWORD BOOKS LIMITED
47 Church Street, Barnsley, South Yorkshire, S70 2AS, England
E-mail: enquiries@pen-and-sword.co.uk
Website: www.pen-and-sword.co.uk

Or
PEN AND SWORD BOOKS
1950 Lawrence Rd, Havertown, PA 19083, USA
E-mail: Uspen-and-sword@casematepublishers.com
Website: www.penandswordbooks.com

CONTENTS

INTRODUCTION

Two men walking on the Moon. Back in 1969 it seemed like something from a science fiction novel. In many ways it still does. Half a century on, the first lunar landing is still regarded by many – conspiracy theorists excepted, of course – as the greatest feat of the twentieth century. And yet, while Apollo 11 will always command a unique place in the history books, it is difficult today to grasp what it meant at the time; how dangerous it was, how much excitement it generated, and how divisive it was.

The mission attracted unprecedented worldwide interest. However, at a time when the USA was grappling with troubles at home and abroad, Apollo 11 polarised opinions. Supporters marvelled at the courage of the astronauts and the remarkable technological advances which made the mission possible. NASA argued that Apollo 11 would benefit all mankind and that even more exciting space adventures would soon follow. Opponents complained that the Moon shot was a waste of money driven by Cold War politics and that the United States should be ashamed of its mixed-up priorities.

It was against this backdrop of unbridled enthusiasm and optimism mixed with bitter controversy that Neil Armstrong, Buzz Aldrin and Michael Collins set off on their historic adventure. So, while this book is the story of man on the Moon, it is also the story of man on Earth in July 1969; a story told not from a twenty-first century perspective, but as it was at the time. It was a time when NASA spoke of landing men on Mars in the 1980s and the US President predicted we would discover life on other planets by the year 2000, when Cold War paranoia meant many Americans feared the Soviet Union might launch an attack from space, and when protestors marched on Cape Kennedy on the eve of the launch demanding action over hunger and poverty. It was a time when the speed of technological advances contrasted starkly with a society in which the First Lady wasn't allowed to watch splashdown because archaic Navy rules barred women from its ships during major operations, and when many African Americans turned their backs on what they regarded as a 'white' mission.

Apollo 11 was man's greatest adventure; an epic 500,000-mile return journey into the unknown. We know how the story ends, but there are many surprises along the way.

CHAPTER 1

RACE TO THE MOON
1957–69

"Let the capitalist countries try to catch up with our country, which has blazed a trail into space."
Soviet Premier Nikita Khrushchev, 1961

"It won't be one man going to the Moon, it will be an entire nation...for all of us must work to put him there."
US President John F. Kennedy, 1961

John F. Kennedy addresses Congress and sets the USA the goal of landing on the Moon by the end of the 1960s.

'RED MOON' SHOCK

4 OCTOBER 1957

The Soviet Union stunned the world today when it launched the first man-made satellite to orbit Earth.

Soviet space scientist Professor Kiril Stanikovich hailed it "a great event for all mankind", but in the USA it has sparked fears that the Communist superpower might launch an attack from space.

Henry Jackson, chairman of the Senate sub-committee for Military Applications of Atomic Energy, said the 'Red Moon' was a "devastating blow" and "a stepping-up of the Cold War".

Major General Holgar Nelson Toftoy, Commander of the Army's Redstone Arsenal in Alabama, claimed that his team, led by former Nazi rocket genius Doctor Wernher Von Braun, could already have put a satellite in orbit had a Navy project not been given priority. Toftoy said: "We said we could do it and, by God, we could. But we were told this wasn't a race."

DOG IN SPACE

3 NOVEMBER 1957

A dog called Laika today became the first live animal in space as the Soviets launched a second successful satellite.

Ivan Alexanderson, Moscow Radio chief, said: "The chief purpose of the project is to prove man's ability to navigate and live in space." The Soviets said medical data showed the dog was in good condition. However, it is understood there's no way of returning it to Earth.

'KAPUTNIK' SHAMBLES

6 DECEMBER 1957

America was left red-faced today when an attempt to launch its first satellite ended in a spectacular live TV failure.

President Dwight Eisenhower has demanded a full report after the Navy-designed Vanguard rocket rose just 4ft at Cape Canaveral, Florida, before exploding. It has been branded 'Kaputnik' and Senate Majority Leader Lyndon Johnson said: "It's one of the most humiliating failures in our history."

AMERICA UP AND AWAY

31 JANUARY 1958

The USA finally has a satellite in orbit after today's successful Explorer 1 launch.

Although the 30lb satellite, tiny in comparison to the Soviet Sputniks, was fired into space on a Juno rocket, a modified version of a Jupiter-C designed by the Army, Vice President Richard Nixon said the USA was committed to "space exploration in the cause of peace".

NEW SPACE AGENCY

1 OCTOBER 1958

A new civilian agency has taken over responsibility for America's space programme.

The National Aeronautics and Space Administration (NASA) came into being today under legislation signed in July by President Dwight Eisenhower. The National Advisory Committee for Aeronautics (NACA) has been absorbed into NASA, along with some Army and Navy staff and facilities.

ASTRONAUTS UNVEILED

9 APRIL 1959

NASA has named seven test pilots as its first team of astronauts.

The men were introduced at a press conference in Washington today. It is hoped one will become the first man in space within two years.

They all underwent rigorous physical and psychological tests, are around 5ft 9in tall so they can fit inside a space capsule, aged in their thirties and married with children. They are Deke Slayton, Scott Carpenter, Gordon Cooper, John Glenn, Gus Grissom, Wally Schirra and Alan Shepard. Slayton said: "I'd give my left arm to be the first man in space."

The USA's first team of astronauts. In the front row are (left to right) Wally Schirra, Deke Slayton, John Glenn and Scott Carpenter and in the back row (left to right) are Alan Shepard, Gus Grissom and Gordon Cooper.

PROBE HITS MOON

14 SEPTEMBER 1959

Soviet Premier Nikita Khrushchev rubbed American noses in it today after the USSR became the first nation to crash-land a craft on the Moon.

Both superpowers have previously sent satellites past the Moon, but Luna 2 is the first flight accurate enough to hit the surface. Khrushchev said: "Only people who deliberately close their eyes and don't want to see reality can doubt the unlimited possibilities of human progress opened by Communism."

FIRST MAN IN SPACE

12 APRIL 1961

America was reeling again today when Yuri Gagarin became the first man in space.

Soviet Air Force pilot Gagarin, 27, orbited the Earth once on Vostok 1 in a 108-minute flight before negotiating re-entry into the atmosphere and parachuting to a safe landing.

The craft flew at 17,000mph – six times faster than any man has ever travelled – and Gagarin reported no problems coping with weightlessness. He told Mission Control in Baikonur, Kazakhstan: "I can observe the Earth. Visibility is good. Everything's working perfectly. I feel fine."

Nikita Khrushchev told the cosmonaut: "You've made yourself immortal. Let the whole world see what our country is capable of. Let the capitalist countries try to catch up with our country, which has blazed a trail into space."

DEBUT FLIGHT DELIGHT

5 MAY 1961

Alan Shepard made America's first manned flight into space today and exclaimed: "Boy, what a ride!"

Launched by a Mercury-Redstone rocket, Shepard, 37, rode Freedom 7 to an altitude of 115 miles before landing in the Atlantic. The sub-orbital flight lasted only fifteen minutes.

1969 MOON DEADLINE

25 MAY 1961

New President John F. Kennedy upped the space race stakes today when he called on America to land a man on the Moon before 1970.

At a special session of Congress, he called for NASA to be backed with $9 billion over the next five years and said: "I believe this nation should commit itself to achieving the goal, before this decade is out, of landing a man on the Moon and returning him safely to the Earth. No single space project in this period will be more impressive to mankind [...] and none will be so difficult or expensive to accomplish."

He added: "In a very real sense, it won't be one man going to the Moon [...] it will be an entire nation. For all of us must work to put him there."

A DAY IN SPACE

6 August 1961

The Soviets have taken another step towards a manned Moon flight after Gherman Titov spent a whole day in space.

Moscow Radio said the cosmonaut, 25, was 'fine' after orbiting the Earth seventeen times. In an unprecedented test of man's ability to cope with long-duration spaceflight, he ate, slept and exercised during the mission.

Nikita Khrushchev radioed him to say: "All the Soviet people are boundlessly happy and are proud of you. Congratulations to you, devoted son of our homeland, of the glorious Communist Party."

GLENN ORBITS EARTH

20 February 1962

America finally put a man in Earth orbit today when John Glenn lapped the planet three times on Friendship 7.

The mission lasted nearly five hours, a huge advance on two brief sub-orbital flights made by Alan Shepard and Gus Grissom. Glenn, 40, survived despite fears that his spacecraft's re-entry heat shield was damaged.

John F. Kennedy said: "We have a long way to go in this space race. This is a new ocean and I believe the United States must sail on it and be in a position second to none. We have more chips on the table than we did some time ago." JFK has written to Nikita Khrushchev proposing the superpowers pool their efforts "before space becomes devoted to uses of war".

NASA LANDING PLAN

11 JULY 1962

American space bosses today finally revealed how they plan to land men on the Moon.

NASA has committed to a method called lunar orbit rendezvous (LOR). Three astronauts will be launched on a huge Saturn V rocket which separates in the early stages of the flight, leaving the crew to journey to the Moon on a command module which houses a separate landing craft. Once in lunar orbit, two of the crew will fly the smaller craft to the surface. They will then blast off and rendezvous with the command module for the return journey.

With Congress backing the programme with billions of dollars, NASA is handing out huge contracts to aerospace companies to design and build hardware. The agency is also building a massive new facility called the Manned Spacecraft Center in Houston.

JFK LUNAR PLEA

12 SEPTEMBER 1962

John F. Kennedy today made an impassioned plea for the American public to support his Moon landing plan.

JFK, whose proposal for a joint lunar mission was rebuffed when he met Nikita Khrushchev in June, told a crowd at Rice University, Houston: "The eyes of the world now look into space, to the Moon and to the planets beyond and we have vowed that we shall not see it governed by a hostile flag of conquest, but by a banner of freedom and peace. We have vowed that we shall not see space filled with weapons of mass destruction."

He added: "Only if the United States occupies a position of pre-eminence can we help decide whether this new ocean will be a sea of peace or a new, terrifying theatre of war. [...] We choose to go to the Moon in this decade and do the other things, not because they're easy, but because they're hard."

COST OF MOON SHOT IS 'NUTS'

12 JUNE 1963

Former US President Dwight Eisenhower today described the prospect of spending up to $30 billion on flying men to the Moon as "nuts".

APOLLO 11: THE MOON LANDING IN REAL TIME

TRANSFER TO LM

SEPARATION OF LM FROM CSM

LANDING ON MOON

FIRST STEP ON MOON

RETURN TO SPACECRAFT

ASCENT STAGE LAUNCH

RENDEZVOUS AND DOCKING

LM JETTISON

NASA illustrations showing how the lunar module will undock from the command module and land on the Moon and how two astronauts will blast off from the surface and rendezvous with the command module for the journey home.

Opposition is growing to John F. Kennedy's scheme and his predecessor weighed into the debate on the day NASA Administrator James Webb formally announced the space agency has ended Project Mercury after six manned flights to focus on the next phase of the programme. Eisenhower complained: "I've never believed that a spectacular dash to the Moon, vastly deepening our debt, is worth the added tax burden it will eventually impose upon our citizens. I suggest that our enthusiasm here be tempered in the interest of fiscal soundness."

The Kennedy administration has kept Congress onside so far, but Eisenhower's outburst highlights the continuing battle JFK and NASA face to secure funding during the two years until the two-man Gemini flights begin and then for Project Apollo. Webb said it would show a "lack of foresight" if the nation backed away from the Moon goal and Vice President Lyndon Johnson insisted: "Last year, the nation bet more on horse races than it spent on space."

TRIBUTE TO KENNEDY

28 NOVEMBER 1963

New President Lyndon Johnson today announced that Cape Canaveral is to be renamed Cape Kennedy in honour of the late John F. Kennedy.

Johnson, who stepped up from Vice President when JFK was assassinated six days ago, told the nation in a live TV address that NASA's facilities in Florida will be called the John F. Kennedy Space Center. The President has pledged to continue Kennedy's lunar landing scheme.

MULTI-CREW FLIGHT

13 OCTOBER 1964

The Soviet Union has chalked up another major achievement – the first multi-crew flight.

Three cosmonauts landed today after orbiting the Earth sixteen times on Voskhod 1 to emphasise the USSR's continued superiority. Last year, Soviet Valentina Tereshkova, 26, became the first woman in space.

America isn't expected to start its multi-crew Gemini flights until next year and James Webb, NASA Administrator, conceded the Soviets' latest first was a "significant accomplishment".

WALKING IN SPACE

18 March 1965

America was upstaged again today when cosmonaut Alexei Leonov made the first spacewalk.

Just five days before NASA makes its first two-man Gemini flight, Leonov, 30, floated outside Voskhod 2 for ten minutes on the end of a tether. Soviet space official Vasily Seleznev confirmed: "The target now is the Moon."

USA RENDEZVOUS BOOST

15 December 1965

The USA is back in the race to the Moon after pulling off the first spacecraft rendezvous.

A pair of two-man craft manoeuvred to within just 6ft of each other in Earth orbit. It's the latest in a string of impressive Gemini missions which have seen America outstrip the Soviets in many areas, including long-duration flights. After the Gemini programme, NASA will begin Project Apollo, a series aimed to culminate with a lunar landing. Lyndon Johnson told the astronauts: "You've moved us one step higher on the stairway to the Moon."

THREE DIE IN US DISASTER

27 January 1967

Three American astronauts died today when Apollo 1 burst into flames during a routine launchpad test.

Gus Grissom, 40, Ed White, 36, and Roger Chaffee, 31, were trapped inside the command module at Cape Kennedy. They were rehearsing for next month's first manned Apollo flight. The programme has been halted pending an inquiry.

James Webb, NASA Administrator, said: "Although everyone realised that someday space pilots would die, who could have thought the first tragedy would be on the ground?"

White's father, Edward, said: "My son died doing a job for his country. We realise these things are inevitable. I'm sure he would have wanted it this way."

COSMONAUT KILLED

24 APRIL 1967

Cosmonaut Vladimir Komarov was killed today when his spaceship crashed into the Earth.

A parachute became tangled as Komarov, 40, returned from orbiting Earth in Soyuz 1, a new type of craft. The mission had ended a lull in the Soviets' manned programme.

APOLLO BACK ON TRACK

22 OCTOBER 1968

NASA is confident it can still meet John F. Kennedy's Moon deadline after finally making the first manned Apollo flight.

The twelve-day Apollo 7 mission, which saw a three-man crew commanded by Wally Schirra test the new craft in Earth orbit, was the USA's first manned flight since the Apollo 1 tragedy twenty months ago put manned missions on hold.

It only went ahead after NASA implemented a raft of safety changes and made unmanned test launches of the Saturn V rocket. The USSR recently orbited an unmanned probe around the Moon and returned it to Earth, but its manned programme has stalled.

AMERICANS ORBIT MOON

24 DECEMBER 1968

The crew of Apollo 8 gave the world an early Christmas present today when they became the first men to fly around the Moon.

A worldwide TV audience watched in wonder as the trio commanded by Frank Borman made ten orbits, beaming back images of the Earth from nearly 250,000 miles away.

In an emotional live broadcast, the astronauts read from Genesis, with Bill Anders opening with: "In the beginning, God created the Heaven and the Earth." Borman signed off with: "Merry Christmas. God bless all of you … all of you on the good Earth."

LUNAR MODULE SUCCESS

13 MARCH 1969

A Moon landing attempt is expected this summer after Apollo 9 successfully tested the lunar module.

In a ten-day mission, Jim McDivitt and Rusty Schweickart flew the bug-like craft in Earth orbit. After today's splashdown, Doctor George Mueller, NASA Associate Administrator for Manned Space Flight, said: "This was as successful a flight as any of us have ever seen."

THE MOON SHOT IS ON

26 MAY 1969

NASA has given the go-ahead for a July lunar landing after a dummy run by Apollo 10 was successfully completed today.

An eight-day mission commanded by Tom Stafford tested every aspect of the Apollo hardware, with the lunar module flying to within nine miles of the Moon's surface.

With the Soviets having only recently managed to dock two spacecraft in Earth orbit, it seems the road is clear for the USA to be first to land men on the Moon.

Doctor Thomas Paine, who became NASA Administrator last year when James Webb retired, confirmed 16 July as the Apollo 11 launch date. Paine pledged: "Eight years ago yesterday, the United States made the decision to land men on the Moon. Today, this moment, we know we *can* go to the Moon. We *will* go to the Moon."

CHAPTER 2

WEDNESDAY, 2 JULY
Launch Minus Fourteen Days

"The lunar landing trip is the most ambitious and daring journey man has attempted during his long struggle from caveman to spaceman."

NASA Rocket Scientist Doctor Wernher Von Braun

"Kennedy's decision to try to put a man on the Moon by the end of the 1960s was rather a wild one."

NASA Administrator Doctor Thomas Paine

*All times are US Eastern Daylight Time unless otherwise stated

Rocket genius Doctor Wernher Von Braun pictured at Cape Kennedy as Apollo 11 is readied for launch.

GOOD TO GO

Apollo 11 was today given the all-clear for its historic Moon flight after a full-scale dress rehearsal for its 16 July launch.

A NASA spokesman said the five-day test at Kennedy Space Center had been "the smoothest thus far in the Apollo series", despite a leaking fuel valve which caused a three-hour delay.

Tomorrow, astronauts Neil Armstrong, Buzz Aldrin and Michael Collins will climb on board their spacecraft on top of the giant $185-million Saturn V – the most powerful rocket ever launched – to practice launch procedures.

The eight-day mission will see Armstrong and Aldrin fly the lunar module to the surface on 20 July. They will spend around two and a half hours walking on the Moon.

The rehearsal ended today with the Saturn fully fuelled with 750,000 gallons of highly-flammable propellant. These include liquid oxygen and hydrogen, which are super-cold fuels, and even the smallest leak causes the liquid to warm and vaporise. Today's leak meant a delay of three hours and eighteen minutes while technicians tightened bolts on a launch tower valve. After the test, the fuel was drained ahead of tomorrow's crew test.

The rocket, with the spacecraft on top of it, was transported to Launchpad 39A two weeks ago. The 363ft-high, 3,200-ton stack was moved from the Vehicle Assembly Building (VAB) three and a half miles away on a flatbed transporter which runs at half a mile an hour on four caterpillar tracks, each weighing a ton.

The 525ft-high VAB is the biggest single-storey building ever constructed. The equivalent of 52 storeys, it could easily house New York's 40-storey UN building and is so big that, before air conditioning was installed, rain clouds formed inside.

'JFK PLEDGE SHOCKED ME'

The space chief handed the job of landing a man on the Moon before 1970 has admitted he was "aghast" at John F. Kennedy's announcement.

Doctor Robert Gilruth was the head of NASA's Space Task Group in 1961 when JFK set the deadline in a speech to Congress. Gilruth recalled: "I could hardly believe my ears. I was literally aghast. America had only fifteen minutes' spaceflight experience. Alan Shepard's flight twenty days earlier had been our first. More than anyone, I realised how much work was required before an American, or any other spaceman, could set foot on the Moon."

He accepted the job because "the challenge was too great" and was appointed Director of the new Manned Spacecraft Center in Houston. The team, working in makeshift offices while the new centre was being built, juggled the ongoing Mercury programme with the Moon project. Gilruth said: "Many of the key ideas evolved in this period of stress, turmoil and major flight activity. We got together evenings, weekends and whenever we could."

Basic spacecraft design and the need for a three-man crew to cope with the mission workload were quickly agreed. When engineers devised the Saturn V, Gilruth had a rocket powerful enough for launch. How to land on the Moon was the issue. Eventually it was decided to have a smaller craft which could separate in lunar orbit, make the landing and then blast off to rejoin the mother ship.

Once Project Mercury finished, attention switched to Gemini. This was a series of Earth-orbit flights to perfect, step by step, the skills for a Moon mission, such as spacecraft docking. Now, after four manned Apollo flights, NASA is set to beat Kennedy's deadline by five months.

$24 BILLION TEAM EFFORT

Industrial leaders and NASA's top boss have hailed America's workforce for making the Moon shot possible.

It has taken $24 billion and a team effort involving thousands of companies and hundreds of thousands of workers. Doctor Thomas Paine, NASA Administrator and the space agency's most senior official, said: "When you think about it, Kennedy's decision to put a man on the Moon by the end of the 1960s was rather a wild one, typical of charismatic leadership. It was a bold act. The programme mobilises the best energies of hundreds of thousands of skilled people. It's a kind of warfare without loss of life; a kind of war effort."

NASA signed up more than a dozen specialist contractors who sub-contracted work to thousands of others. In 1966, when work was at its peak, it involved more than 20,000 businesses and 400,000 people.

Countless challenges have included devising stronger, lighter, smaller and more reliable materials and parts.

North American Rockwell is the biggest single contractor with a $3.6 billion deal for the command and service modules and Saturn V second stage. William Bergen, Space Division President, said: "This is the finest industrial team ever assembled. I look for the effects to continue long after the landing and to be felt in every facet of industry."

His colleague Donn Williams, Aerospace Senior Vice President, added: "The people in England in the 1700s must have thought the early settlers in America were crazy for leaving the comforts of England for the unknown wilds of a new continent. People must have thought the same about American pioneers who pushed west. That's the way it is when it comes to space. I'm sure the rewards are there."

Joseph Gavin, a Vice President with $1.8-billion lunar module prime contractor Grumman Aerospace, said: "The thing that's impressed me most about the programme has been the unflagging individual enthusiasm demonstrated day and night."

Other major contractors include Boeing, with a $1.5 billion deal for the Saturn's first stage, and McDonnell Douglas, paid $1 billion for the third stage.

A LITTLE PIECE OF HISTORY

Hundreds of thousands of Americans will nervously follow the mission, ready to claim their own little place in history.

Excitement is building at NASA's various centres and at thousands of private companies.

NASA's Houston centre, home of Mission Control, has been at the hub of the effort since opening in 1963. Andrew Sea, who has worked there since day one, said: "This is a once-in-a-lifetime thing, like being on Christopher Columbus' crew. All these years we've been working and it's coming true. What an adventure!"

Charles Bauer, of the Special Events team, explained: "People like to feel they're part of the Moon programme. Even a secretary can say 'The astronauts couldn't have done it without my help'."

The 1,620-acre site houses around 4,000 government and 9,000 private workers. The average salary for the government staff is $15,400, but Thomas Conger, Facilities Planner, said: "We're not so much working for the salary, but for the psychic income. There's something special about being close to this Moon effort."

In Florida, two thirds of people living near the Kennedy Space Center are involved in the programme. Ray Forbes, of contractor General Electric, is a regular visitor and explained: "When you come down here you over-smoke, over-eat, over-drink, over-worry and under-sleep. When you leave, you get a feeling of depression because all this adrenalin has been pumping."

Among other key NASA facilities are the Goddard Space Flight Center in Greenbelt, Maryland, and Marshall Space Flight Center in Huntsville, Alabama. The Saturn V was developed in Huntsville and Goddard is the hub of a $600-million worldwide communications network featuring land, air, sea and space tracking stations. Doctor John Clark, Goddard Director, said: "We're the ears of Mission Control."

Typical of the many private-sector workers involved in the programme is Bill Rooney, a solderer and one of 6,000 Grumman staff who built the lunar module. He said: "If something should go wrong, I'll worry it was something I did. Could I have done it better?"

Rooney, 35, added: "You work on that thing for so long, it seems like a monster. But you work a little longer and it's a beautiful thing. The overtime money alone wasn't worth it – not seeing your family until after the kids are asleep, not having a normal social life. Except, of course, it's the Moon shot."

ARMSTRONG'S LUCKY BREAK

Neil Armstrong insists his role as the first man to walk on the Moon is down to luck.

NASA has taken a methodical, step-by-step approach towards the Moon shot and it's only because of the near-faultless performances of Apollo missions so far that it falls on Armstrong and Buzz Aldrin to make the first landing attempt.

Armstrong is keen to play down his individual role and, in a pre-flight interview with *Life* magazine, he said: "I really do hope the public will recognise this as a national effort. Who the individual happens to be who takes the first step on the Moon is just the way things fell."

The Apollo 11 crew was named in January, before it was certain that it would be the first mission to attempt a landing. Deke Slayton, the astronauts' chief who selects the crews, said: "There's no way you could sit down two or three years ago and say 'These are the guys who'll be first on the Moon'. It was about luck as much as anything. It could just as easily have been someone on the next flight or the one before. It depended on how the cards fell."

Armstrong insists the enormity of his role hasn't sunk in. "I don't think much about the emotional aspects. Every time I've flown a new airplane or spacecraft, I've looked around and said 'By George, I'm really flying it!'. I suspect the same thoughts will come on the Moon... 'By George, I'm really here!'"

Unlike serving Air Force officers Buzz Aldrin and Michael Collins, Armstrong is a civilian astronaut, although he was a Navy pilot from 1949 to 1952, flying missions in the Korean War.

After securing an aeronautical engineering degree, he joined NACA in 1955 as an engineer and pilot for experimental aircraft. In 1962 he was one of the second group of astronauts selected by NASA and is renowned for his cool head, as demonstrated on his previous mission as Gemini 8 Commander in 1966.

When Buzz Aldrin (left), Neil Armstrong (centre) and Michael Collins were unveiled in January as the Apollo 11 crew, they weren't certain their mission would be to attempt a first lunar landing.

He and Dave Scott made the first docking with another craft – the unmanned Agena – but a faulty thruster pitched them into a violent tumble. The men were in danger of losing consciousness, but Armstrong undocked and regained control.

Armstrong also cheated death last year when the Lunar Landing Research Vehicle – a vertical take-off training contraption dubbed the 'Flying Bedstead' – lost propulsion and he parachuted to safety before it crashed and exploded.

The astronaut and wife Jan have two sons aged twelve and six, but seven years ago their second child, Karen, tragically died aged just two as a result of a brain tumour.

NO-NONSENSE ALDRIN

Buzz Aldrin says he's risking his life for the benefit of science rather than personal glory.

The Air Force Colonel, 39, will earn his own place in history when he becomes the second man to walk on the Moon. Speaking to *Life* magazine, the married father-of-three said: "There's little risk if everything goes according to plan, but nobody can guarantee it. But the risk isn't the lure. I see my part less in a romantic and personal way and more as a contribution to increasing knowledge and understanding for future generations."

Ted Guillory, a NASA Flightplan Writer, describes Aldrin as "the best scientific mind we've sent into space". He's an expert on spacecraft rendezvous, having written his astronautics degree thesis on the subject. Colleagues call him 'Doctor Rendezvous' and Guillory said: "Boy, he's really something! I sometimes think he could correct a computer."

Named Edwin, after his father, he has been known as Buzz since childhood because his sister mispronounced 'brother' as 'buzzer'. His dad was a distinguished aviation pioneer and Buzz inherited his passion for flying, becoming a fighter pilot after graduating from West Point Military Academy. He received the Distinguished Flying Cross in the Korean War.

Having studied at Massachusetts Institute of Technology, he was in NASA's third batch of astronauts in 1963. Three years later he flew on Gemini 12 and spent a record total of five and a half hours spacewalking.

He was perhaps destined for Apollo 11. His late mother's maiden name was Moon.

OFF TO THE MOON ... 'YIPPEE!'

Michael Collins can't wait for the mission after missing out on an earlier piece of history.

Collins, 38, missed the Apollo 8 Moon flight after a shoulder operation. Jim Lovell stepped in and Collins was told his next flight would be with the other members of the Apollo 8 back-up crew: Neil Armstrong and Buzz Aldrin.

Now set to fly to the Moon after all, the married father-of-three told *Life* magazine: "It made me very sad to get bumped off a great flight. Then, before they named the crew for the first landing, I heard rumours I was going to be on it. Finally the word came and I felt like ... 'Whoosh! Yippee!'."

As Command Module Pilot, he'll orbit the Moon alone during the landing. He said: "People ask how I feel about going down in history, but I don't really think I will. That may be more for Neil and Buzz."

A West Point graduate and serving Air Force officer, the Lieutenant Colonel was born in Italy when his dad was stationed there with the Army. In 1952, after gaining a science degree, Collins joined the Air Force and later became a test pilot. He joined NASA in 1963 and flew on Gemini 10 in 1966, making a spacewalk.

John Glenn's spaceflight inspired him to join NASA. "I really hadn't thought too much about space before that. I mean, it was something limited to Buck Rogers and science fiction!"

FROM CAVEMAN TO SPACEMAN

The rocket genius who has gone from designing Nazi weapons to leading the US space effort says Apollo 11 will confirm man's evolution "from caveman to spaceman".

Doctor Wernher Von Braun spearheaded Germany's World War II rocket programme before surrendering to the Allies in 1945 and moving to the USA with more than a hundred of his scientists. Now Director of Marshall Space Flight Center (MSFC), he made Moon flights possible by developing the Saturn V.

He has always said his team were reluctant participants in the Nazi war effort and that space was their real goal. In an article written for United Press, he said: "The lunar landing is the most ambitious and daring journey man has attempted during his long struggle from caveman to spaceman. It marks the beginning of a new era in the extension of his domain. It's a thrill for all mankind, the fulfilment of one of our oldest dreams.

"Besides the sheer technological skill involved, the most impressive thing is the organisation of men and women with diverse skills and talents, scattered across the country, into a unit working toward a single purpose. There are many problems around us, many of them crying out for a solution. Can we not apply the same determination and ability towards solving problems on Earth?"

Von Braun helped develop long-range weapons like the V2, which Adolf Hitler used to terrorise Britain. After arriving in America his team worked with the Army, but Von Braun pushed for their talents to be turned to space, even making TV shows with Walt Disney to arouse public interest.

His expertise made America's early satellite launches possible and, when the Redstone Arsenal in Alabama became the MSFC under the newly-formed NASA, he was put in charge.

'MONSTROUS' ROCKET

An engineer who helped design the giant Moon rocket has admitted he wasn't sure it would ever fly.

The Saturn V is 60ft taller than the Statue of Liberty and Doctor William Mrazek, one of the German rocket team who came to the USA after World War II, said: "Many of us have more than three decades of experience in rocketry, but when we completed the first drawings of the Saturn V, we couldn't believe this monstrous machine would ever lift itself off the launchpad."

Its engines have combined horsepower equivalent to 548 jet fighters and it creates 7,600,000lb of thrust at launch; eighty-five times more power than the Hoover Dam. Mrazek explained: "When Kennedy called for a trip to the Moon, it was an entirely new ball game. This nation's first satellite, Explorer 1, weighed 30lb. This super-rocket was intended to orbit the weight of more than nine thousand Explorer 1s."

The answer was a three-stage rocket which burns enough fuel for a 30 miles-per-gallon family car to drive nine million miles. Each stage is discarded after performing its function. Five F1 engines power stage one, hence the Roman numeral in the rocket's name. This stage enables lift-off and builds the spacecraft's speed to 6,430mph before being jettisoned two minutes and forty-two seconds into the flight. Stage two's five J2 engines take over, burning for about six and a half minutes. When the second stage separates, stage three's single J2 engine is fired for about two and a half minutes to take Apollo into Earth orbit at a speed of 17,432mph and an altitude of about 120 miles. Two hours and thirty-two minutes into the flight, the stage three engine is ignited again for around six minutes to boost the spacecraft out of orbit at 24,000mph and on a course for the Moon. After separation, the command module turns around to retrieve the lunar module – garaged inside the third stage – before heading on.

Of the original 117-strong team of German scientists, fifty-four still work together. Mrazek said: "Basically, we're doing the same jobs today we did then. We had something no-one else had. Everyone fitted like a mosaic stone; if one dropped out, we felt it."

CRUNCH TIME FOR 'UGLY BEAST'

The Moon landing's success rests on a spacecraft even its own designers admit looks like it was knocked up by a DIY enthusiast.

The lunar module faces its ultimate test when Neil Armstrong and Buzz Aldrin attempt to land on the Moon and, a day later, take off again.

Grumman Aerospace was tasked with producing the unique craft and the result was a flimsy, insect-like machine. Stephen Tsontakis, a senior Grumman engineer, said: "It's an ugly beast by layman's standards. Only an engineer would appreciate the sophisticated design."

The project was so complex that Grumman delivered the craft two years late at a cost of $1.8 billion, four times the original estimate. Joseph Gavin, who had overall financial responsibility for the project, said: "A lot of people may look at it and say 'I could have done that myself in the cellar', but they forget there was no instruction sheet. We had to figure it out for ourselves."

The craft has two million functioning parts – compared to 3,000 in an average family car – and weight was a big problem. Even the Saturn V has a limit on what it can lift and, with fuel accounting for three-quarters of the module's maximum weight, drastic measures included removing the seats. The pilots will stand up, with Velcro keeping them on their feet in near-zero gravity. The module has two engines, one for descent and one for blasting off. The descent stage acts as a makeshift launchpad.

The Apollo 9 lunar module is flown in Earth orbit. Now the 'ugly beast' faces its ultimate test.

Insulation was another issue. The Sun will bake one side at 320 degrees Fahrenheit, while the other side is in the shade at minus 250. The craft looks like it's wrapped in kitchen foil and Grumman engineer Moe Tawil said: "It's really like a Thermos bottle. The insulation is so efficient it lets out less heat than a 100-watt lightbulb."

WE COME IN PEACE

A plaque declaring mankind's peaceful intentions will be left on the Moon, NASA revealed today.

Attached to the lunar module's descent stage, it features the names and signatures of the crew and US President Richard Nixon and is inscribed with the message: 'Here men from the planet Earth first set foot upon the Moon. July 1969 AD. We came in peace for all mankind.'

A hi-tech silicon disc bearing goodwill messages from world leaders will also be left behind. It's only one and a half inches in diameter, but the process used to make micro-miniature electronic circuits has been used to etch on the words. These appear as tiny dots, but can be read using a microscope.

A Stars & Stripes flag, with built-in tubing to support it in the airless atmosphere, will be raised on an aluminium flagstaff. Miniature flags from 136 other nations and US states will be taken in the lunar module and returned to be presented as gifts.

The official Apollo 11 logo has also been unveiled. It shows an eagle with an olive branch in its talons landing on the Moon, with the Earth rising in the distance. In a tradition first set by Gemini 5, the astronauts came up with the design and this one is based on sketches made by Michael Collins.

ASTRONAUTS' $1 MILLION JACKPOT

Neil Armstrong will be paid about $30 for the time he spends walking on the Moon, but America's astronaut team could bank $1 million.

That's the amount that spin-offs from the mission are expected to generate over the next twelve months. Under a NASA pool system, fifty-five astronauts and the widows of eight others will get an equal share.

Astronauts are paid a relatively modest amount. As a civilian, Armstrong gets $27,401 a year, so will earn $33 for the Moonwalk. Buzz Aldrin and Michael Collins receive Air Force salaries: Aldrin, a Colonel, earns $18,623; Collins, a Lieutenant Colonel, earns $17,148. All three get benefits, including a house in Houston.

One astronaut team bonus is a deal with publisher Time-Life, which has had rights to exclusive interviews since 1959. This will bring in $200,000 this year, plus a likely $500,000 from syndication rights paid for reproducing *Life* interviews. Time-Life's latest deal includes plans for a book, plus interviews on vinyl and tape. NASA's astronauts can expect $250,000 from the book and $100,000 from audio sales.

The Time-Life arrangement is controversial, but Paul Haney, former NASA Public Affairs Director, explained: "One of the fundamental reasons was the $100,000 life assurance contract each man received.

After the Apollo 1 fire, the widows received the money within seventy-two hours." The government was unable to provide such substantial life cover.

Criticism has led to revisions in recent deals, with exclusivity no longer applying to astronauts talking about missions and training. Instead, *Life* has the rights to interviews about their personal lives.

NIXON'S MEET & GREET

Richard Nixon will greet the Moon men minutes after they splash down in the Pacific, the White House announced today.

The President will fly out to watch from the recovery ship, USS *Hornet*, before meeting the astronauts. Nixon will then head off on a mini-tour of Asia.

The new President has been accused of milking the Moon shot to boost his own image. On Monday he and his wife, Pat, dined with Apollo 10 heroes Tom Stafford, Gene Cernan and John Young. They presented him with four framed US flags and Stafford told him: "These flags have been thirty-one times around the Moon. We thought you'd like them just the way we brought them back. That's why we didn't press out the wrinkles."

EARLIER MOONWALK PLEA

NASA chiefs are under pressure to stage the Moonwalk earlier to suit US TV schedules.

When Neil Armstrong steps on to surface, it will be 11.12pm on 20 July on the west coast and 2.12am on 21 July in the east.

Networks have spent millions on coverage and are disappointed the crowning moment won't be primetime viewing. The landing has been arranged for lunar dawn so that, with the Sun behind them, Armstrong and Buzz Aldrin will be able to see potential hazards. There will be a ten-hour delay until the Moonwalk, but TV bosses hope it will be brought forward by NASA scrapping a rest period.

Some parts of the world, particularly South America, could miss live launch coverage after a satellite failure last week led to pictures of the investiture of Britain's Prince Charles as Prince of Wales having to be re-routed. Another communications satellite is already in place, but using this to plug the gap in TV coverage could interfere with mission transmissions.

MONKEY IN ORBIT

The astronauts won't be alone in space – they'll have a monkey called Bonny for company.

The three-year-old male pigtail monkey is already into his fourth day in Earth orbit in a $92-million mission. He is due to spend thirty days in space – a record for a live animal – to help pave the way for longer manned flights.

Doctor William Ross Adey, of the University of California, said: "This mission should yield more biological data than all of the manned flights so far put together."

Bonny was sleeping when the rocket blasted off and his heart rate jumped to more than 200bpm, but Doctor Charles Wilson, Project Manager, reported today: "Bonny's in good shape. He slept well after the first shock and he's eating well and drinking plenty of water."

Biosatellite 3 is the third in a series of missions to study the effects of weightlessness, cosmic radiation and other factors. Previous flights carried plants and insects. The longest manned spaceflight was the fourteen-day Gemini 7 mission in 1965, which indicated that weightlessness causes temporary blood, bone and muscle changes in humans. When, in 1966, the Soviets orbited two dogs for twenty-two days, they suffered liver malfunctions.

Bonny is strapped into a flying lab with numerous sensors attached to him. His arms have been left free to push buttons in behavioural tests designed to monitor his mental state. If he punches the correct buttons, food pellets pop out of a dispenser.

He has steel wires embedded in his head for scientists to make the first detailed studies of brain activity in weightlessness. NASA said today it had "seen clear evidence of Bonny having dreamed during sleep".

US HERO'S SOVIET VISIT

Apollo 8 hero Frank Borman today became the first American astronaut to visit the Soviet Union.

Borman arrived in Moscow with his wife and two teenage sons after their flight suffered twenty hours of delays, including a stop in Newfoundland because of engine trouble. When he was greeted by cosmonauts Georgi Beregovoi, Konstantin Feoktistov and Gherman

Titov at Sheremyetovo Airport, Borman joked: "It took almost as long to get from New York to Moscow as it did from the Earth to the Moon!"

Although arranged by the Soviet Society for Friendship, it is thought the ten-day visit has had state approval. The Air Force Colonel hopes he may be allowed to visit the secretive Baikonur Cosmodrome. Ex-French President Charles de Gaulle is the only Westerner ever to go there.

Borman quit active astronaut duty earlier this year and is now Field Director of NASA's Space Station Task Group. The Soviets have rejected numerous invitations for cosmonauts to visit Cape Kennedy, but Borman said: "I'm sure that'll be discussed."

CHAPTER 3

THURSDAY, 3 JULY
Launch Minus Thirteen days

"Everybody is aware we're going to the Moon, but there are no jitters."

> Launch Director Rocco Petrone

"Hundreds of thousands of problems could keep Apollo 11 from landing on the Moon."

> Apollo 10 Astronaut Gene Cernan

The Apollo 11 crew leave the launchpad after today's successful launch rehearsal.

WAS BUZZ BUMPED?

NASA has denied Neil Armstrong pulled rank to ensure he makes history.

Armstrong will step on to the surface about twenty minutes before Buzz Aldrin, but former NASA Public Affairs Director Paul Haney today claimed it was originally planned that the Lunar Module Pilot would go first.

Haney said: "I do know it caused quite an upset. Neil's a very competitive fellow. I think he wasn't unaware of the importance of the first man who stepped on to the Moon, looked at it very carefully and decided it should be Commander's prerogative."

Deke Slayton, Director of Flight Crew Operations, responded by insisting it was a management decision with no input from the astronauts. He said Armstrong has a clearer route to the exit and Aldrin will spend less time on the surface to conserve energy for key tasks in the spacecraft.

Haney quit NASA earlier this year after being told he was being switched to a desk job in Washington. He now works for ITN, a UK TV company.

CREW'S FAMILY WEEKEND

The astronauts flew to Houston tonight for a final break with their families after a successful launch rehearsal at Cape Kennedy.

The crew donned full spacesuits for the launchpad test on board the command module on top of the Saturn V rocket. It followed a five-day dress rehearsal of countdown, including fuelling the rocket.

Rocco Petrone, Launch Director, said: "It all went off extremely smoothly. Everybody's aware we're going to the Moon, but there are no jitters, no feeling that this is any more difficult than past launches."

The astronauts followed the exact timetable for the real launch from the moment they were woken at 4.45am. Once on board, they ran through checks on communications, instruments and life-support systems. For safety reasons, the rocket had been drained of fuel.

NO PIECE OF CAKE

Apollo 10 astronaut Gene Cernan has cautioned against complacency over the Moon landing.

Cernan and Tom Stafford successfully flew their lunar module close to the Moon's surface in May, but did have one major scare. At one point the craft briefly tumbled out of control and a worldwide audience heard

Cernan exclaim: "Son of a bitch! What the hell happened?" It's thought the problem was caused by an incorrect switch setting.

Today, Cernan warned that the latest mission won't be "a piece of cake". He said: "They could encounter any of the problems we encountered plus – and I don't hesitate to say it – hundreds of thousands of other problems that could keep Apollo 11 from landing on the Moon."

DATE FUDGED

A commemorative plaque to be left on the Moon deliberately doesn't feature the exact date of the landing.

The plaque's inscription has the date July 1969, but not specifically 20 July. A NASA spokesman explained today: "If we claim to be representing all mankind, we really can't include the date. Although the touchdown will be on July 20 in America it will, for example, be July 21 in Australia.

NASA has released this illustration showing the inscriptions on a plaque which will be left on the Moon.

"There's another reason. If the launch is delayed, we wouldn't have time to replace the plaque. We felt it was best to leave the inscription somewhat vague."

MOON MENU

Bacon, beef stew and cookies will be among the first items of food ever eaten on the Moon.

The menu on offer to astronauts has come a long way since John Glenn made do with squeezing a tube of pureed beef and vegetables into his mouth.

When Neil Armstrong and Buzz Aldrin tuck into two meals on the lunar surface, they'll enjoy the benefits of huge advances in space food technology. Apollo 11 has the most varied pantry of any flight and Doctor Malcolm Smith, Chief of Food and Nutrition, has gone to great lengths to improve the menu, even interviewing astronauts' wives about their husbands' eating habits.

He said: "Astronauts complain, but when you ask them what it is they don't like, all they can say is 'It's atrocious'. Like most men, they aren't fussy eaters; they'll eat what you put in front of them."

NASA knows it's vital that astronauts maintain good energy levels, but Smith added: "The more tension there has been in the flights, the less the crews have eaten. They become so involved with operations that they go long stretches without eating. The Apollo 9 crew ate nothing for twenty-eight hours. All our efforts for good nutrition mean nothing if they don't eat."

The daily calorie target per astronaut is 2,800 and Smith balances this against offering more choice. Over the past decade, scientists have had to work out how to preserve food without refrigeration and to overcome the problem of weightlessness. The advent of freeze-drying, as used for instant coffee, was one major breakthrough.

The astronauts will have a mixture of meal types. Some are rehydrated with water – such as soups and puddings sucked out of a bag through a tube – and others are ready to eat. Some meals can now be eaten with a spoon after it was found that foods with a certain moisture content don't float away. Sandwiches are available thanks to non-crumbling bread and the newest innovations are foil-wrapped 'wet pack' meals, such as turkey and gravy, which don't need rehydrating.

Smith admitted the improved menu had one drawback. He explained: "One of the penalties the crew are paying is that, as we add more natural foods – which aren't necessarily low in residue – they create a bigger problem for our primitive waste-management system."

Excreta is stored in plastic bags, with germicidal pellets used to prevent bacterial growth and odours. Urine is collected using rubber tubes fastened to the astronauts and is dumped into space.

Medical Director Doctor Charles Berry says questions about the toilet facilities "are the ones I'm most frequently asked by heads of state and people on the street".

He added: "The management systems for both liquid and solid wastes are of very questionable acceptability. They serve the purpose but are rather primitive."

'WE'LL FLY TOGETHER'

America and the Soviet Union will make joint space missions by the mid-1970s, Frank Borman predicted today.

The astronaut, in Leningrad on the first full day of his ten-day visit to the USSR, forecast the USA would have a station in Earth orbit within six years and said: "I foresee a time when Soviet and American spacemen will be flying together."

He was enthusiastically welcomed in Leningrad, with Mayor Nikolai Sizov telling him: "Any spaceman, American or Soviet, is a beloved guest here."

Back in Houston, meanwhile, speculation is growing that the Soviets plan an attempt to upstage America by sending an unmanned probe to the Moon and return lunar samples to Earth before Apollo 11 does. Cosmonaut Alexei Leonov said recently that the Soviets hoped to have Moon samples on display at the World Fair in Japan next year.

TAXI TO THE MOON

A NASA official today outlined plans to have an Earth-Moon 'taxi' service running within a decade.

Doctor George Mueller, Associate Administrator for Manned Space Flight, said the reusable shuttle would need multi-billion-dollar funding at "roughly the cost of bringing Apollo into being".

Speaking at Canaveral Press Club, he said: "The real question facing the nation now is making a series of decisions in short order if we're to have a continuing space programme. This scheme would really open up space for mankind."

The three-stage shuttle would transfer astronauts between the ground and an Earth-orbit station, ferry them to and from a second station in lunar orbit and then between there and the Moon. Mueller said the system would be more cost-effective in the long-term and that research into developing a reusable rocket, possibly nuclear-powered, was under way.

RACE FOR SPACES

Florida faces its own race for space – finding enough beds for a million tourists.

That's how many people Cocoa Beach Chamber of Commerce expects to descend on the area around Cape Kennedy for launch and, with hotel rooms booked up three months ago, alternative schemes are being put in place. Residents are being asked to take in guests and beach camps are being set up.

The closest any visitor will get to the launch site is Port Canaveral, about five miles away. The Lemon River between Cocoa Beach and Merritt Island is expected to be cluttered with boats as people try to get an unhindered view.

'OLD GLORY' FLIES HIGH

American space successes have helped spark an unprecedented spike in Stars & Stripes sales.

Flag manufacturers are reporting record sales and tens of millions are flying on cars, storefronts and homes.

The Vietnam War and domestic unrest have hit morale in recent years, but Apollo has helped restore many people's national pride. Charles Williams, of Atlanta flag-maker ABC Supply, said sales of 'Old Glory' had increased by a third this year. He said: "People are tired of internal upheaval, civil strife. They want to get back to honesty, to the Stars & Stripes."

Planting the flag on the lunar surface is a thorny issue. John F. Kennedy said the Moon must not be "governed by a hostile flag of conquest, but by a banner of freedom and peace" and the Stars & Stripes has been a visible symbol during US missions.

However, the 1967 'Outer Space Treaty' bars territorial claims or military activity and Richard Nixon stressed in his inaugural address in January that celestial bodies should be treated "not as new worlds to be conquered, but as a new adventure to be shared".

A NASA committee considered whether a UN flag should be raised instead, but concluded the Stars & Stripes was appropriate to represent where the astronauts came from. It was decided, though, that miniature flags from all UN members should be carried to and from the Moon and that a plaque should be left stressing the landing was 'for all mankind'.

In fact, Republican Richard Roudebush recently succeeded in adding a controversial amendment to a NASA funding bill barring any US mission wholly paid for by Congress from planting any flag other than America's.

CHAPTER 4

FRIDAY, 4 JULY
Launch Minus Twelve Days

"We'll get to the Moon, all right – you will find out when."
Soviet Cosmonaut Georgy Beregovoy

"Is it worthwhile? Is it worthwhile to lift the spirits of millions of human beings? If not, what is worthwhile?"
Presidential Science Adviser Doctor Lee DuBridge

A view of the top of the Apollo 11 rocket stack showing the White Room through which the astronauts will enter the command module ahead of launch.

SANITISED PRESS CONFERENCE

Extraordinary measures will be taken to keep the astronauts germ-free when they face the media in Houston tomorrow.

In their final public appearance, Neil Armstrong, Buzz Aldrin and Michael Collins will sit in a $5,718 open-faced plastic tent to help protect them from germs. Reporters will be 50ft away and powerful air blowers will keep the air the media breathe away from the stage.

After the hour-long press conference they will give one-on-one interviews via an intercom system from behind air-tight glass walls at the Lunar Receiving Laboratory, the facility where they will be quarantined for nearly three weeks after the mission.

The astronauts spent today at home celebrating Independence Day. Tomorrow, as well as their media duties, they will work in flight simulators in Houston. On Sunday, they will have a last day with their families before returning to Florida on Monday.

Doctor Robert Gilruth, Manned Space Flight Director, is satisfied everything is in place for their final preparations. He said: "We certainly haven't any corners. This crew is knowledgeable and well trained. The ground crews are in good shape, too. I certainly can't imagine how we could be any more ready."

SATURN AND GLOSS

Engineers were today busy giving the Moon rocket an emergency paint job.

Thermal paint peeled off the Saturn V's upper stage when it was filled with super-cooled fuel during the launch rehearsal.

Lab tests suggest it was down to the wrong type of primer being used rather than any problem with the rocket, but the repainting is vital. Early in the flight, it deflects heat to ensure the fuel which will fire the spacecraft out of Earth orbit is kept cool.

SPIRITED DEFENCE

Richard Nixon's Science Adviser today made a rousing defence of the space programme.

NASA is poised for back-to-back successes, with the lunar landing followed within weeks by two unmanned probes making a fly-by of Mars.

Speaking at Independence Day celebrations in Dearborn, Michigan, Doctor Lee DuBridge hit back at those who claim it is wrong to pump money into space exploration when so many social problems need addressing.

DuBridge said: "For untold millions of years, the human animal was chained to the Earth. This month, the first man will set foot on another world. Later this month, two spacecraft will reach Mars and send back new information about that planet. Americans will have no reason to be ashamed on those days. Is it worthwhile? Is it worthwhile to lift the spirits of millions of human beings? If not, what else *is* worthwhile?

"The social problems which beset us aren't all that easy. In this area, human beings aren't working together, but are in conflict. We don't yet know the cause of these troubles nor do we have the mechanisms for curing them. We must try and we'll often fail. We'll learn from our failures."

The Mars probes should provide the most detailed pictures ever of the 'Red Planet', revealing new information about its atmosphere, composition and water content. Mariner 6, launched on 24 February, has travelled forty-one million miles and is due to reach Mars on 31 July. Mariner 7, which followed on 27 March, will also fly within 2,000 miles of the surface on 5 August.

'WE WON'T SEE GOD IN SPACE'

Saturn V mastermind Doctor Wernher Von Braun has hit back at those who claim space exploration has undermined religious beliefs.

Nikita Khrushchev used early Soviet manned spaceflights to ridicule those who believe in heaven, while many atheists suggest man's escape from the Earth's atmosphere proves that God doesn't exist.

But in a revealing interview about his personal beliefs, former Nazi rocket engineer Von Braun insists space exploration should strengthen people's faith.

Von Braun, who has become a devout Christian while living in America, said he found it "as difficult to understand a scientist who doesn't acknowledge the presence of a superior rationality behind the existence of the universe as it is to comprehend a theologian who would deny the advances of science".

He said: "Manned spaceflight has opened for us, thus far, only a tiny door for viewing the awesome reaches of space. Our outlook through this peephole at the vast mysteries of the universe only confirms our belief in its creator."

After the USSR's first two manned spaceflights, Khrushchev said: "We Communists don't believe in after-life. As for paradise, we've heard a lot about it from priests, so we decided to find out for ourselves. We sent up our explorer, Yuri Gagarin. He circled the globe and found nothing in outer space. It's pitch dark there, he said. No Garden of Eden, nothing like heaven.

"So we sent Gherman Titov and told him to fly for a whole day. After all, Gagarin was up there only an hour and a half, so he might have missed paradise. We told Titov to take a good look. Well, he confirmed Gagarin's conclusion. He reported there was nothing there."

But Von Braun insisted: "There's certainly no scientific reason why God can't retain the same position in our modern world that he held before we began probing his creation. Any effort to visualise God, to reduce him to our comprehension, beggars his greatness. Astronomy and space exploration are teaching us that the good Lord is the master of a greater kingdom."

DESTINY OF A DREAMER

Neil Armstrong's high school science teacher has revealed how the astronaut told him as a 16-year-old that he hoped to fly to the Moon one day.

John Crites taught the teenage Armstrong in Wapakoneta, Ohio, and is among guests personally invited to the launch by the astronaut.

Armstrong was a standout student and Crites recalled: "I considered Neil a dreamer, and I don't mean that to be uncomplimentary. He wanted to do something daring and different. He always had a goal to work on. I recall him saying to me 'Someday I'd like to go the Moon'."

Crites revealed that, in his letter inviting him to the launch, the astronaut wrote: "We've been fortunate to live in this particular page of history and I've been uniquely lucky to participate."

Wapakoneta amateur astronomer Jacob Zint also remembers Armstrong's interest in the Moon. Zint, who had a telescope in his backyard, said: "Most of the kids would look for two minutes and that would be enough, but Neil would look and look. He was always asking if I thought there was life out there. I do believe the Moon was Neil's point of interest right from the beginning."

BATTLE OVER LANDING PLAN

A space engineer has revealed how he fought for two years to convince NASA to adopt his plan for the Moon landing.

The mission might not be happening had Doctor John Houbolt, who was an aeronautical engineer at NASA's Langley Research Center in Virginia, not won the argument for lunar orbit rendezvous (LOR).

NASA initially planned a return journey with a huge rocket called the Nova. It then considered Earth orbit rendezvous (EOR), in which two separate Saturn V rockets would have been launched, with the lunar spacecraft being assembled and fuelled in Earth orbit.

But Houbolt realised LOR needed less fuel, that the initial vehicle launched from Earth could be smaller and that the whole operation would be cheaper and easier.

He said: "When you approach the Moon in the main spacecraft – your living room – why take the whole living room down to the surface? Why not leave it in orbit, descend in a small landing craft, then re-join the main living quarters by means of a rendezvous? I thought 'This is fantastic.' I vowed to dedicate myself to the task. But no-one was interested.

"I presented the idea many times. The reaction to the idea of having astronauts perform a rendezvous 240,000 miles from the Earth was generally negative; in some cases direct hostility. Even Wernher Von Braun originally thought it was too hairy."

Finally, though, he began to win over opinion and in July 1962 – with Von Braun now onside – NASA adopted LOR. "Suddenly, after two years of indifference, there was practically complete agreement."

LIFE-OR-DEATH DECISIONS

NASA'S top doctor has spoken of the personal burden he carries in making potential life-or-death mission decisions.

The Moon landing is the latest journey into the unknown since the USA's first manned flight and Doctor Charles Berry, Medical Director, has faced many dilemmas during that time.

He said: "We've knocked down one medical fear after another. I'm confident man is going to adapt to this environment. Before Alan Shepard's Mercury flight in 1961 there were many dire predictions about what might happen; he'd become disorientated, mentally unbalanced, things like that. The President's Science Advisory Committee wanted to stop Shepard's flight. They were convinced that launching a man on a rocket would produce such high heart rates he'd end up with cardiac failure."

In 1965 Jim McDivitt and Ed White spent an unprecedented four days in space and Berry recalled: "I received phone calls in the control centre almost up to the time of launch from other responsible people who said in effect 'You're going to kill these men'. That was a momentous flight that knocked down a lot of straw men."

Berry's team will constantly monitor medical data during the Moonwalk. If necessary, the astronauts will be told to rest or even abort. Berry said: "We'll watch them closely to see how fast they tire. What we learn will be very important to planning future lunar explorations."

The 4,514 hours US astronauts have spent in weightlessness have thrown up some medical issues. As the Mercury missions became longer, astronauts' hearts became lazy because their bodies found it easier to circulate blood in a weightless environment. Introducing exercises during missions has reduced this cardiovascular deconditioning.

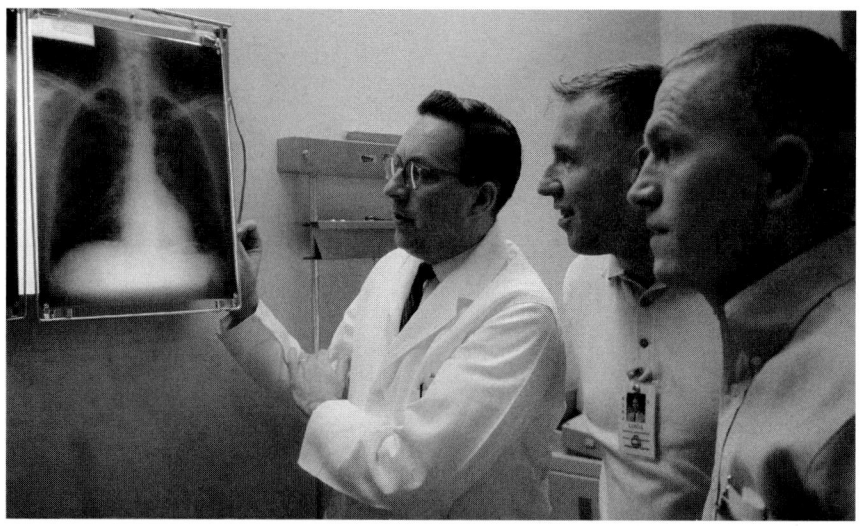

Doctor Charles Berry (left) pictured with astronauts Jim Lovell (centre) and Frank Borman at a pre-flight examination in 1965. Berry has faced countless new medical challenges during the space programme.

Longer-duration Gemini flights led to anaemia because of a loss of red blood cells. When safety changes were brought in after Apollo 1 – with the spacecraft cabin being filled with a mix of oxygen and nitrogen rather than 100 per cent oxygen – Berry's team realised the loss of red blood cells was caused by a pure oxygen atmosphere.

The closest Berry came to losing a crew in flight was in 1966 when Neil Armstrong and Dave Scott's Gemini craft went into an uncontrollable spin during a docking test. Berry admitted: "They were rapidly approaching the limits of disorientation. Another few seconds and we might have lost them."

Despite the pressures, Berry said: "It's a wonderful thing, working in the future. It's a real medical frontier and I see things no-one has ever seen before."

A NEW LANGUAGE

Words and phrases inspired by the Moon shot are set to find their way into dictionaries.

'Space speak' is in increasingly common use and Massachusetts-based G&C Merriam, which has been producing dictionaries since 1843, says some examples are being considered for inclusion in its next edition.

Words including 'Moontel' – a lunar hotel – and 'lunarnaut', meaning an astronaut who has been to the Moon, could feature. Doctor Philip Gove, Editorial Team Leader, said: "Most new words get their greatest push into everyday language from newspapers and TV. War and social upheaval have the most dramatic impact on our language."

An incredible 352 acronyms and abbreviations feature in the Apollo 11 flight plan and the astronauts' wives have had to learn this new language. Pat Collins said: "I didn't want to live in a vacuum of ignorance as my husband advanced."

COSMONAUTS' SECRET

Cosmonauts who attended a US Independence Day reception in Moscow today were staying tight-lipped over a possible Soviet Moon shot.

Speculation is rife that an unmanned probe will be launched in a bid to bring lunar samples back to Earth before Apollo 11 does. Several cosmonauts were guests at an event hosted by US Ambassador Jacob

Beam, but wouldn't be drawn over the possible mission. Georgy Beregovoy, involved last year in the first Soviet space rendezvous, said: "We'll get to the Moon all right … you'll find out when."

Astronaut Frank Borman was at the reception at which Soviet guests – including two high-ranking government ministers – were able to try pizza and hot dogs. In a goodwill gesture on the second full day of his visit to the USSR, Borman suggested the Soviets send two boys of a similar age to his sons Frederick, 17, and Edwin, 15, for a vacation in the USA with his family. He joked: "We promise not to make hippies out of them!"

SPACE JUNK HITS SHIP

Japanese delegates to the UN today claimed a ship sailing off the coast of Siberia was hit by falling debris from a Soviet spacecraft.

They said the freighter *Dai Chi Chinei* was damaged and five sailors seriously injured on 5 June in the first known case of 'space junk' causing damage on Earth. The Japanese claim two Soviet ships collected most of the wreckage, but pieces the freighter captain retrieved were found to be from a Soviet spacecraft.

The Japanese revealed the details in a bid to break the deadlock in negotiations over procedures for liability in such cases, with delegate Shigeo Iwai saying it proved the discussions were "more than merely theoretical".

BONNY GO-SLOW

Space monkey Bonny was today given the go-ahead to stay in Earth orbit, despite signs he is on a go-slow.

Although Bonny is said to be in good condition after six days on Biosatellite 3, a NASA spokesman admitted: "He's shown a disinterest in performing his tasks."

Bonny can earn up to forty extra food pellets by carrying out behavioural tasks correctly, but yesterday ate only twenty-nine and today that dropped to twenty-two. He also drank less water.

Doctor William Ross Adey, California University Brain Unit Research Director, said: "The animal is going very rapidly from one state of sleep to another. He's been going from being wide awake to a deep sleep in thirty seconds."

CHAPTER 5

SATURDAY, 5 JULY
Launch Minus Eleven Days

"Everything conceivable has been done to assure our safety. But we must realise things can go wrong."
> *Apollo 11 Commander Neil Armstrong*

"If they have difficulty on the Moon there's nothing I can do. They know and I know and Mission Control knows I simply light the motor and come home without them."
> *Apollo 11 Command Module Pilot Michael Collins*

Neil Armstrong (left), Buzz Aldrin (centre) and Michael Collins at today's press conference.

'WE COULD DIE ON MOON'

Neil Armstrong today admitted he and Buzz Aldrin could be stranded on the Moon.

The Apollo 11 crew made their last public appearance today at a press conference in Houston.

Pressed by reporters over the perils of the mission, Armstrong conceded that rescue would be impossible if the lunar module failed to take off and that he and Aldrin would be doomed to die when their oxygen ran out.

Armstrong said: "It's an unpleasant thing we've chosen not to think about. We don't think that's a likely situation, but certainly it's a possible one. Everything conceivable has been done to ensure our safety, but we must realise things can go wrong."

Michael Collins said: "If they have difficulty on the Moon, there's nothing I can do. I guess the question everyone has is how do I feel about having to leave them? I don't think that'll happen. If it did, I'd do everything to help. But they know and I know and Mission Control knows there are certain categories of malfunctions where I simply light the motor and come home without them."

During the Apollo 10 mission, the lunar module briefly span out of control and this has raised concerns whether it's up to the job. But Armstrong insisted: "We have high confidence in our systems. Otherwise we wouldn't go."

Asked about the landing, he confirmed he's likely to take manual control in the final stages. He said: "We'll have a view of the landing area from about two or three minutes before landing. We'll be continuously observing it for a smooth touchdown point and an absence of large craters. If required, we'll alter that landing point. We'll probably have a minute and a half or so of fuel which will permit us to go perhaps 1,000ft if we choose."

Aldrin admitted there's no guarantee they'll stay after landing. He said: "One of the first things we'll be engaged in is making decisions as to how long we'll be staying on the surface. There are several abort points. We'll be checking various systems to get a stay or no-stay decision."

Armstrong confirmed they may choose to scrap a rest period and start the Moonwalk early. "We'd certainly like to reserve that possibility. On flights in the past, I guess, we've demonstrated there's a somewhat more-than-average ability to change our minds and shift things around."

Armstrong claimed he hadn't thought about what he'll say when he steps on to the Moon, but said of his historic assignment: "I suppose the chance of doing what thousands of people through history have wanted to do is beyond the ability of the odds-maker to calculate. For me to fulfil the dream is a stroke of incredibly good fortune."

He dismissed claims he used his position as Commander to bump Aldrin into second spot for the Moonwalk, saying: "My recommendation was never asked nor given."

On comparisons with Christopher Columbus' voyages of discovery, Armstrong said: "Our flight varies from many of the great explorations in that they probed into great areas of the unknown. We think we know a good bit about where we're going. All in all, ours isn't a flight into the unknown."

The astronauts arrived on stage in Houston wearing masks to protect them from germs, but took them off during the session. Asked where he'd like to go on vacation after the mission, Armstrong joked: "The Lunar Receiving Laboratory. If I'm able to go there, it means we'll have succeeded."

'THIRD MAN' NOT FRUSTRATED

Michael Collins insists he isn't "the slightest bit frustrated" that he won't get to walk on the Moon.

Collins faces twenty-eight hours orbiting alone while his colleagues make their expedition and has already been dubbed the 'forgotten man' by the media.

But he told today's press conference: "I'm going 99.9 per cent of the way. That suits me fine. I'd probably be less than honest if I said I had the best seat of the three. On the other hand, I can also say with complete honesty that I'm happy. They couldn't be doing what they're doing unless I were there doing my third."

He raised a laugh when asked why a pre-flight ban on contact with anyone outside the immediate NASA team didn't extend to the astronauts' families. "My wife and children have signed a statement that they have no germs!"

Collins admitted he was uneasy about the pre-flight hype, saying: "I mean, we're still here! We haven't been anywhere, we haven't done anything and there's all this tremendous hubbub. Maybe we ought to wait until the flight's over and see whether this grand and glorious thing works out."

CODENAMES REVEALED

The codenames for the Apollo 11 spacecraft are 'Columbia' and 'Eagle', it was revealed today.

The command module's name is a reference to the District of Columbia – home to Congress – and a nod to Jules Verne's 101-year-old novel about a Moon flight on the spacecraft Columbiad. Eagle was chosen for the lunar module because it's a national symbol.

Neil Armstrong said: "We give the spacecraft names so you don't confuse them when listening to transmissions. Those names were suggested by a number of people. They're representative of the flight and of the nation's hopes."

Asked about planting the Stars & Stripes on the Moon, he said: "That decision has been made where it should be made, in Congress. I wouldn't presume to question it."

HOP-ALONG BUZZ

Buzz Aldrin says people can expect to see him and Neil Armstrong hopping like kangaroos on the Moon.

The pair have practiced with trips in a jet transport plane, with the pilot flying up and down, rollercoaster-style, to simulate weightlessness in thirty-second bursts. Aldrin told today's press conference it was "quite easy" to move at a "fairly rapid pace". He said: "There have been several techniques employed. It looks like you can walk conventionally, one foot after another. It also looks as though you can do a two-footed hop, kangaroo-style."

The pair don't expect to venture further than 70ft from the lunar module, but Armstrong said: "If there was an item of unusual interest that would warrant inspection at the expense of our planned tasks, we'd certainly want the freedom to make that inspection. I can't say what such an object would be. If it were an animal, I'm sure we'd go and have a look; or maybe we'd go the other way!"

NIXON WARNED OFF

Richard Nixon was today urged to cancel an eve-of-launch dinner with the astronauts.

Doctor Charles Berry, NASA Medical Director, is worried the President could pass on germs and said Nixon would have to sit downwind of powerful air blowers as a precaution. He said: "I have to practice good preventative medicine regardless and I have to say the dinner is the sort of thing I'd discourage."

When the astronauts splash down on 24 July, Nixon will greet them on the recovery ship. Although they will speak via intercom through an air-tight window, Berry said Nixon had been advised there was still a risk he could end up in quarantine if there was accidental direct contact.

Nixon, meanwhile, is facing accusations of using the mission for personal political gain. His signature features on a plaque to be left on the Moon and today's *Washington Post* editorial said: "The names of the three brave men who made the voyage would seem to us to be enough."

Britain's *Sunday Telegraph* commented that Nixon appeared "determined to use the Moon shot as a launchpad for a world tour that smacks of public relations gimmickry. The mask of the new world statesman has slipped to reveal the old 'Tricky Dicky'".

LAUNCH TEAM FIRED UP

The former Army officer in charge of the launch has warned that the astronauts will be in danger if his team produce anything less than perfection.

Every manned Apollo launch, starting with Apollo 7, has gone perfectly and an Apollo 11 countdown rehearsal went off with barely a hitch.

But Rocco Petrone, Launch Director, said today: "Despite marvellous computer techniques, the three men on top of the rocket are riding on the capability of the 500 men in the firing room. They're very dedicated people who know their jobs. Their operation has to be a combination of the discipline and proficiency of a professional football team and a military unit.

"It's in the last twenty-two minutes that the countdown gets really dynamic. That's when the high-pressure hydrogen starts flowing into the tank bottles and, from then on, the guys are watching everything like hawks. Those last twenty-two minutes are when we really earn our pay."

Communications were so poor for early launches that the firing room was in a blast-proof concrete bunker next to the launchpad. Now Petrone directs launches from three and a half miles away. He can make radio contact with anyone in the firing room or on the launchpad and has constant chatter coming through his headset. "I can talk to any one of

Doctor Wernher Von Braun (left) and Launch Director Rocco Petrone.

the hundreds of men on the team, but the twenty channels I have tuned in all the time are the ones I'm likely to have to deal with most frequently."

Launch Director since 1966, Petrone oversaw the planning and building of the launch facilities. He has responsibility for the mission until the rocket lifts off but will then hand over to Houston, where four Flight Directors – Clifford Charlesworth, Gerald Griffin, Eugene Kranz and Glynn Lunney – will take turns in charge.

LATE-NIGHT VIEWING

Nearly half of Americans who own a TV set plan to watch the Moonwalk.

A survey conducted for US network ABC suggests tens of millions won't be put off even if it happens in the early hours. There are about eighty-one million TV sets in the USA and the poll predicts that around forty-three million will be tuned in at a time when normally only eight per cent of sets are in use. George Hage, Apollo 11 Mission Director, predicted: "There's going to be a lot of lost sleep that night!"

A billion people worldwide are expected to watch live or recorded footage of the Moonwalk, with an even bigger figure expected for radio. At least a third of the population won't see a single moment of the mission. TVs are a rarity in less-developed nations such as India, where there are only 8,500 for a population of 530 million, and the 762 million people of Communist China are expected to face a news blackout on the mission. It is also unclear if Soviets will see any footage.

In South Africa, the government has resisted the introduction of TV, claiming it would be a bad influence on society. Mission footage will be shown in cinemas several days after it happens. Some wealthy South Africans are beating the ban by paying $800 a head for a five-night trip to England, including a chartered flight and a London hotel room with a TV set.

The public will see live colour coverage of launch and splashdown, plus seven live in-flight telecasts. There will be no live pictures of the landing, but the Moonwalk will be shown via cameras deployed after Eagle lands. This footage will be in black and white because of technical issues involved in receiving a colour signal from the Moon's surface.

If the signal isn't strong enough, the astronauts will set up an umbrella-shaped antenna. Chester Lee, Assistant Mission Director, said: "This would take ten minutes and it would impact the entire timeline we've laid out for the astronauts, but we're prepared to do it."

BORMAN'S LENIN TRIBUTE

Frank Borman today paid a public tribute to Vladimir Lenin and three Soviet space heroes.

The astronaut had a hectic day in Moscow, including a tour of the Star City cosmonaut base and a visit to the Kremlin, where he and his wife, Susan, placed red carnations on the Kremlin Wall graves of space pioneers Yuri Gagarin, Sergei Korolev and Vladimir Komarov.

Gargarin died in a plane crash last year and Komarov was killed when his spacecraft crashed on re-entry in 1967. Korolev was the rocket genius behind early Soviet successes, but his role was kept secret until his death in 1966.

Borman placed wreaths at the mausoleum where revolutionary leader Lenin is buried and at the Soviet Tomb of the Unknown Soldier.

This afternoon he became the first US astronaut to visit the forest base where the cosmonauts live and work. Radio Moscow said the Bormans received a "warm reception" and that they were presented with a model of Gagarin's Vostok 1.

WORRIES OVER MONKEY

Scientists were today keeping a wary eye on Bonny the space monkey on the seventh day of his planned thirty-day flight on Biosatellite 3.

Bonny's metabolism has been slowing down for several days and speculation is mounting that the mission may even have to be aborted. Doctor Chris Wilson, Project Manager, said: "The monkey hasn't been converting as much food to energy and isn't very active. He continues to show disinterest in performing his behavioural tasks."

CHAPTER 6

SUNDAY, 6 JULY
Launch Minus Ten Days

"Neil and Buzz are much better than just amateur geologists. They won't be like someone picking up interesting or pretty shells at the seashore."

NASA Geologist Doctor John Dietrich

"There were very few space toys on the market two years ago. By next year they'll be coming out of our ears."

Advertising Agency Executive William Silverstein

Buzz Aldrin (left) and Neil Armstrong hone their geology skills during a field trip in Texas in February.

CREW'S LAST DAY OFF

The astronauts enjoyed a day off in Houston today, the last private time they'll spend with their families for five weeks.

Tomorrow they'll fly back to Florida and won't be at home again until completing a quarantine period lasting twenty-one days from the moment they leave the Moon.

The strains of the public spotlight were apparent in comments made during yesterday's press day. Asked if their lives will ever be the same again, Neil Armstrong admitted: "If there's any disadvantage to being in the position I am in, that's it."

Buzz Aldrin said: "We may anticipate, at the moment, an unusual amount of notoriety, but I'm not convinced this will be of a lasting nature. This is something that does go with the mission. My family look on this as a tremendous challenge for me and also as an invasion, somewhat, of their privacy."

He said his three children – aged between eleven and thirteen – had "all of a sudden realised their father is involved in this endeavour of national importance".

Explaining how his wife, Joan, copes, Aldrin added: "The profession I was engaged in at the time we were married, that of test pilot, was certainly a dangerous one. She fully understood the dangers. I don't think we consider the dangers now are any greater; they certainly are more spectacular."

Michael Collins, who has three children aged between six and ten, said: "They're really too young to appreciate the implications. I got them together and tried to explain the historic import. They sort of listened and nodded. There was only one question from the youngest …'Who's driving?'."

He added: "I like to live a normal private life. My intent is to take whatever action's necessary to continue that. The world's divided into two kinds of people; those who want more public exposure and those who want less. I'm just one of those who prefers less."

The astronauts live in suburban communities in Houston. Armstrong lives in El Lago, with Collins and Aldrin a ten-minute drive away in Nassau Bay.

Jan Armstrong met her husband at university in Indiana while studying home economics. She's a keen swimmer who coaches the local synchronised swimming team. Joan Aldrin – who met Buzz at a dinner party and courted him largely by post when he was away with the Air Force – is an amateur actress with a dramatic art and literature degree. Pat Collins met her husband while doing a civilian job at a US Air Force base in France.

ALDRIN'S PRIVATE CHURCH SERVICE

Devout Christian Buzz Aldrin made sure NASA's strict pre-flight quarantine arrangements didn't force him to miss church today.

Aldrin is an elder and lay reader at the Presbyterian Church in Webster, Texas. The astronauts are under strict orders not to have face-to-face contact with strangers, but he was given clearance for a private communion service with Reverend Dean Woodruff once the Sunday congregation left.

YOU GUYS ROCK

Neil Armstrong and Buzz Aldrin have been given an A-plus for their geology 'homework'.

After months of training to ensure they know what to look for when collecting lunar samples, Doctor Ted Foss, Lunar Geology Director, says they have become "tremendously capable".

He said: "Even if geologists chose who they wanted for this mission, they'd have a hard time finding a better pair. They have a great interest, because they realise how important it is."

Training has included field trips to places ranging from the Grand Canyon to Iceland. Doctor John Dietrich, a NASA geologist, said: "They're much better than just amateur geologists. They won't be like someone picking up interesting or pretty shells at the seashore. They'll know what they're doing. They've absorbed all that has been thrown at them and have nearly photographic memories.

"They'll be talking to us, describing what they see on the surface. We may be able to suggest they pick up particular rocks, but they'll be in a better position to judge. They'll be like a football player who has to make split-second decisions in action. The coach can have something to say later about what he should have done, but not at that moment. On field trips, we deliberately salted the area with exotic rocks from another region on Earth, but they've been able to pick out these rocks as foreign. They have sharp powers of observation."

Doctor Elbert King, Curator of NASA's Lunar Receiving Laboratory, hopes the samples will reveal how old the Moon is and how it was formed, possibly settling arguments about whether craters were caused by volcanic activity or meteor strikes, or both. King said: "A lunar rock will be like an Aladdin's lamp. If you rub it with the right instrument, it'll tell you secrets of the universe."

On future missions, NASA plans to include astronauts selected specifically for their scientific backgrounds. These include geologist Doctor Jack Schmitt, who has helped train Armstrong and Aldrin.

COMPUTER WORLD

The space age is opening up a new frontier on Earth – the world of computers.

A raft of technological advances have made the Moon shot possible, but Doctor Chris Kraft, NASA Director of Flight Operations, is in no doubt which has been the most significant.

He said: "If I had to single out the piece of equipment that, more than any other, has allowed us to go from Earth-orbit flights to lunar trips in just over seven years, it would be the high-speed computer. The progress made since the Mercury days is staggering. It's worth a large portion of the dollar outlay for the space programme. It has given us a wealth of knowledge that can be applied to almost every engineering, scientific and industrial endeavour. Today, we can get almost instantaneous data. The doctor on duty can study the electrocardiogram from the astronauts as though they were patients in his office."

Computers link the spacecraft, Cape Kennedy, Mission Control and a worldwide tracking network which includes Goddard Space Flight Center, seventeen ground stations and a dozen ships and planes.

The final four and a half hours of countdown are completely computer-controlled. Throughout the flight, computers will relay and analyse spacecraft data, with the 'fourth brain' on board Apollo 11 keeping track of speed and position, calculating course changes and looking for system malfunctions. Similar technology will help Neil Armstrong and Buzz Aldrin land on the Moon.

Leroy Hall, of NASA's Flight Computer Branch, said: "Without computers, you could never do the job; you need computations instantly. Slide rules wouldn't be accurate enough."

The 6Mb programme developed by IBM to run computers at Cape Kennedy and Houston is the most complex software ever written. Houston's computers for Mercury missions had a storage capacity of 32,000 words. Today, Mission Control's main computer has a capacity of five and a half million words. On the command module, the astronauts control the craft by typing simple coded instructions into a computer which has 64kb of RAM and operates at 0.043MHz.

Although computers are vital, Doctor George Mueller, Associate Administrator for Manned Space Flight, is at pains to stress humans' key role: "There's a popular fad among intellectuals to decry mankind. There's no machine that begins to equal the abilities of man's brain. I haven't noticed that the number of artists has decreased because the number of photographers has increased."

TOYS ON LAUNCHPAD

Toy companies are hoping for a space lift-off at the second attempt.

Despite space games and toys flopping in the late 1950s and early 1960s, manufacturers are flooding the market with a new batch of products timed to cash in on Apollo 11.

The industry is targeting retail sales of $2.7 billion in 1969 in the USA alone. According to New York-based financial researcher Value Line: "Toy-makers are counting on the success of the Apollo programme to breathe new life into space toys."

William Silverstein, President of the Adams Dana Silverstein advertising agency, predicted: "There were very few space items on the market two years ago. By next year they'll be coming out of our ears!"

Lionel Weintraub, President of the Toy Manufacturers of America, said: "Until last year, space wasn't particularly popular with children. This year, there's been a noticeable improvement in the climate."

Robert Silverstein, Eldon Industries President, said: "Space toys weren't accepted by children three or four years ago. Today, children relate to space. It's part of the environment they live in."

Toys on the market range from action figures and model spacecraft to games such as Parker Bros' *Situation 7* and ES Loew's *Count Down*.

Some old favourites have been given a space makeover, such as Gabriel Industries' *Gilbert Erector Set*, which has been around for fifty-six years. New kits include a lunar module and a robot.

Even cartoon favourite *Snoopy* has enjoyed an unlikely boost after the Apollo 10 crew named their command module Charlie Brown and their lunar module Snoopy. That sparked an upsurge in sales of merchandise related to Charles M. Schultz's *Peanuts* cartoon strip.

BORMAN VISITS TRACKING STATION

Frank Borman today got another peak behind the Soviet space curtain.

The astronaut visited one of the USSR's main space tracking stations in the Crimea. His family weren't allowed to go and stayed behind in Yalta, a Black Sea resort on the 'Russian Riviera'.

Earlier, the Bormans visited a 4,500-capacity camp for Young Pioneers, a Communist youth group. They were given a demonstration of Russian-style folk dancing – the Soviet equivalent of a hoe-down – and after initially protesting he was "too clumsy" to take part, the astronaut joined in.

'AFRONAUTS' PREPARE FOR SPACE

An African space pioneer says he will press ahead with training a dozen would-be astronauts, despite his failure to secure funding for a Moon mission.

Edward Makuka Nkoloso, Zambian Academy of Space Research chief, has been working on his programme since 1960 and dreams of sending what he calls his 'Afronauts' into space. He claims to have technological ideas "years ahead" of the Americans and Soviets, but his bids for funding from the two superpowers and the UN have been rebuffed.

His cadets' training includes being rolled down hills inside 44-gallon drums to simulate weightlessness. For splashdown, they are suspended in barrels above a pond and the rope is cut. Nkoloso said: "Some people think I'm crazy, but I'll be laughing the day I plant Zambia's flag on the Moon."

CHAPTER 7

MONDAY, 7 JULY
Launch Minus Nine Days

"A society that can appropriate vast billions to put man in a place where it was considered only God could reach deserves acclaim for achievement and contempt for bizarre social values."
                    Civil Rights Leader Reverend Ralph Abernathy

"I'd be willing to eat lunar samples ... to expose myself and my family to Moon rocks."
                    NASA Scientist Doctor William Kemmerer

Michael Collins, cup of coffee in hand, arrives for work today at the Kennedy Space Center flight crew training building.

BACK TO BUSINESS

The astronauts knuckled down to their final preparations today after flying back to Florida from a final break with their families in Houston.

Neil Armstrong indicated in Saturday's press conference that they hoped to shorten their working days in the lead-up to launch, saying: "We'd sure like to do that. I wouldn't compromise our requirement to be ready to fly – and that will take precedence – but we'd like to attempt to knock down to something like a twelve to fourteen-hour workday and get a reasonable amount of sleep each night."

But today they worked late into the night in simulators, with a run-through of the lunar blast-off and procedures for lunar orbit rendezvous.

NIXON U-TURN

Richard Nixon today scrapped plans to dine with the astronauts the night before launch.

The U-turn came after NASA's medical chief, Doctor Charles Berry, spoke out against the plan. Berry claimed the President might pass on germs.

White House Press Secretary Ronald Ziegler said the decision was taken because of "additional factors NASA feels enter into this". He confirmed Nixon's trip to watch splashdown *will* go ahead, despite Berry warning there was a slight risk he could end up in quarantine.

'MOON BUGS' DEBATE

Space agency chiefs have played down fears that harmful 'Moon bugs' could be brought back to Earth.

The debate over possible contamination has been fuelled by *The Andromeda Strain*, a new novel by Harvard medical graduate Doctor Michael Crichton, in which America is devastated by a virus carried to Earth by a satellite.

Doctor Martin Alexander, Professor of Soil Microbiology at Cornell University, New York, believes NASA's precautions are "totally inadequate" and that the crew's quarantine period is too short.

Astrobiologist Carl Sagan, who has worked as an adviser to NASA, warned: "Maybe it's sure to ninety-nine per cent that Apollo 11 won't

The Apollo 11 astronauts, wearing biological suits and masks, during a rehearsal for splashdown recovery earlier this year. Some scientists fear opening the command module hatch at sea could release dangerous lunar organisms.

bring back lunar organisms, but even that one per cent of uncertainty is too large to be complacent about."

Doctor Persa Bell, Director of the $8 million Lunar Receiving Laboratory (LRL), promised no chances would be taken. He said: "Our lab will have a quarantine more severe than those imposed at the strictest communicable disease hospitals in the world."

Doctor William Kemmerer, the lab's Manager for Medicine, is so confident he pledged: "I'd be willing to eat lunar samples … be willing to expose myself and my family to Moon rocks. However, the quarantine process is valid. No-one should release any lunar sample until he's sure it's not dangerous."

All procedures have been approved by a committee made up of experts from NASA, the National Academy of Science and various government departments. NASA says safety measures will begin as soon as Neil Armstrong and Buzz Aldrin leave the lunar surface, with the pair vacuuming each other before re-entering the command module.

At splashdown, frogmen wearing special protective gear will open the spacecraft hatch and the astronauts, wearing biological suits and breathing masks, will be airlifted on to the USS *Hornet*. Divers will close the hatch and sterilise the capsule before it is lifted aboard.

The astronauts will be immediately isolated inside the Mobile Quarantine Facility (MQF), an air-tight trailer, and the spacecraft will be sealed off separately. After arriving in Hawaii, the MQF will be flown to Houston and the astronauts will be transferred to the LRL to continue quarantine. Doctor Charles Berry, Medical Director, said: "Three weeks … the astronauts aren't going to like it. But they're highly motivated and tremendously responsible men. They know they have to take it."

Lunar samples will be brought back from the Moon in air-tight boxes. These will be sealed in vacuum chambers and mice will be exposed to the material before scientists handle it.

PINPOINT ACCURACY

The mastermind behind Apollo's hi-tech guidance system says he convinced space chiefs that lunar navigation wasn't impossible by telling them he'd happily fly the spacecraft himself.

NASA turned to Doctor Charles Draper, from Massachusetts Institute of Technology (MIT), to tackle the seemingly insurmountable navigational challenges posed by a 500,000-mile return trip to the Moon.

Draper, an expert on gyroscopic instrumentation, was adamant it could be done. Recalling his early discussions with NASA, he said: "I told them I'd go along and run it myself. As I remember it, this ended the interview!"

Early flights were almost entirely remotely guided from Mission Control, but this isn't possible for Apollo; not least because there is a communications blackout when a spacecraft is on the far side of the Moon. Contractors led by General Motors helped MIT design and build the command module equipment, which is about the size of a suitcase.

An inertial measurement unit (IMU), containing gyroscopes and accelerometers, works out the spacecraft's position and attitude (angle). The gyroscopes, spinning wheels in a moveable frame, help the craft recognise up from down, left from right and backward from forward. The crew update the IMU by making star sightings using a sextant – a traditional hand-held navigational instrument – and feeding the results into an optical unit. A miniaturised computer makes continuous calculations based on the information provided by the IMU and optical unit.

Even a tiny miscalculation would send Apollo massively off course, but Draper believes the chances of errors are virtually zero and, on previous Apollo flights, the system has worked perfectly.

CRASH LANDING

Journalists invited to try landing the lunar module found out today just how tough it will be.

NASA allowed a group of reporters to have a go in a simulator at Massachusetts Institute of Technology (MIT), where boffins devised technology used in the real module.

The journalists chalked up more aborts and crashes than touchdowns as they found themselves bewildered by the number of controls, alarms and flashing lights in the cockpit.

MIT engineers have spent countless hours in simulators rehearsing the landing. Edwin Olsson, from NASA's MIT Instrumentation Lab, said: "The purpose has been to check the programmes under stress and to make sure you don't overload the computer."

The lunar module computer was built by Raytheon and has 64Kb of RAM. Only core data stays in the memory, with the system constantly erasing and rewriting information such as the craft's location.

Massachusetts-based Raytheon built systems used in Polaris missiles, but the Apollo job was so complex that, within a year, the firm had increased the number of staff dedicated to it from 800 to 2,000.

As important as they are, even computers have limits. Neil Armstrong will take manual control in the final stages of landing and Olsson admitted: "The computer won't pick out boulders for you!"

DON'T EASE UP, SAYS NASA CHIEF

NASA'S top official today warned that the Soviets will soar back ahead in the space race if America takes its foot off the gas.

A task force will make recommendations to Richard Nixon on 1 September on future plans. Earth and lunar-orbit space stations are under consideration, as well as a manned Mars mission.

Doctor Thomas Paine, NASA Administrator, said: "The lunar landing will be the culmination of America satisfying everyone that it is indeed the leading nation it thought it was before Sputnik blazed across the skies. There's always the danger that we may feel we can now slacken off. If we weren't to start new programmes now, the situation might well reverse and the Soviets might once again develop superior technological capabilities in space."

'SHAME ON AMERICA'

A civil rights campaigner has poured scorn on the "bizarre social values" behind the $24 billion space programme.

Reverend Ralph Abernathy, President of the Southern Christian Leadership Conference, claims the nation should be ashamed of putting the Moon shot ahead of social issues.

Abernathy, who was a close friend and associate of the late Doctor Martin Luther King, said: "A society that can resolve to conquer space, to put man in a place where in ages past it was considered only God could reach, to appropriate vast billions, to systematically set about to discover the necessary scientific knowledge … that society deserves both acclaim and contempt.

"Acclaim for achievement and contempt for bizarre social values. For, though it has the capacity to meet extraordinary challenges, it has failed to use its ability to rid itself of the scourges of racism, poverty and war, all of which are brutally scarring the nation."

Abernathy is one of several opinion-makers interviewed in the latest edition of *Newsweek* magazine. Another, American philosopher Lewis Mumford, claimed: "Space exploration is strictly a military by-product and, without pressure from the Pentagon and Kremlin, it would never have found a place in any national budget."

SOVIETS' APOLLO SNUB

More than 2,500 journalists from forty-six countries will cover the mission, but none from the Soviet Union.

Of 520 overseas media accredited by NASA so far, Japan has the most, with 104. Other well-represented nations include Italy (sixty-one), Britain (forty-eight), France (thirty-nine), West Germany (thirty-six), Mexico (thirty-three), Spain (eighteen), Canada (seventeen), Australia (sixteen), Argentina (fifteen), Switzerland (thirteen), Brazil (twelve) and Belgium (ten).

Of communist countries, Yugoslavia and Romania are represented, but there have been no applications from the USSR or China.

The massive media presence will be a stark contrast to the early days of US rocket launches at what was then called the Cape Canaveral Missile Annex. The Department of Defense staged tests amid tight security and reporters relied on local gossip to find out when they were happening and would watch through binoculars from the balconies of their Cocoa Beach motels.

MONKEY MISSION ABORTED

A dramatic bid to rescue space monkey Bonny was under way tonight after NASA aborted his mission.

The plug has been pulled nine days into the thirty-day flight amid concern over the pigtail monkey's deteriorating condition.

Doctor Charles Wilson, Project Manager, said: "During the night, the primate refused water and appeared to be in a deeper state of rest than at other times in his flight. An attempt was made to alert him by repeated 'water available' signals, a signal to which this animal normally responds.

"His lack of response was interpreted to indicate a sluggishness which, if allowed to continue, could have led to serious deterioration."

Data showed Bonny was sleeping for long periods, had a steadily-lowering body temperature, a reduced heart rate and shallow breathing.

The Air Force will attempt to recover Biosatellite 3 in mid-air over the Pacific tomorrow after it re-enters the atmosphere. A C119 'Flying Boxcar' should be able to snag the capsule's parachute and reel in the craft like a fish on a line. If not, it will be plucked from the water.

CHAPTER 8

TUESDAY, 8 JULY
Launch Minus Eight Days

"One shouldn't be too surprised if Apollo 11 failed, if Armstrong and Aldrin were to die. A boulder or a disguised crevasse could turn triumph to disaster."

Space Expert Professor Stuart Butler

"The Moon is sacred to our people and we predict that man won't land on the Moon."

Native American Indian Chief Joseph Logan Junior

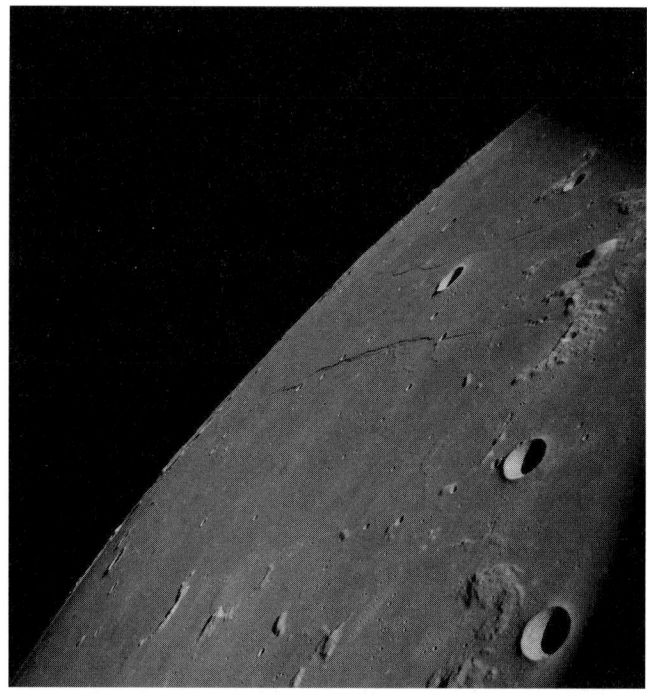

A general view of the Sea of Tranquility, taken during the Apollo 8 mission, where Apollo 11 will land.

PERIL 'PLAYED DOWN'

Space agency bosses were today accused of glossing over the chances of Neil Armstrong and Buzz Aldrin dying on the Moon.

In a no-holds-barred newspaper article in his native Australia, world-renowned physicist and space expert Professor Stuart Butler spelt out the perils of the mission.

Butler, Professor of Theoretical Physics at the University of Sydney, claimed: "American space experts have tended to underplay the risks they'll face during the unrehearsed journey into the unknown. But perhaps Christopher Kraft, of NASA, spelled it out when he said recently 'I don't know what the odds are that we'll achieve a landing at the first attempt; maybe three to one'.

"The Americans have shown that every phase can be just about guaranteed as safe, save the actual lunar descent and ascent. Once the controller decides all systems are go, Armstrong and Aldrin are on their own in an area foreign to man. That's the point of no return. They'll be flying into a danger zone that nobody, despite the intensity and thoroughness of preparations, can fully neutralise."

Describing in Melbourne newspaper *The Age* how difficult the landing will be, he explained: "Two men will be attempting to land a rocket backwards on its exhaust flame on unknown terrain. They'll have to settle it down on its tripod legs, exactly upright. If it lands too hard, it will bounce and perhaps topple over. If the surface is uneven, it will land askew and perhaps topple.

"Imagine the following. You are 50,000ft above the surface. You fire the rocket. You begin losing altitude. Your accuracy must never vary by more than a fraction of a degree. You must not hit at too high a speed. You must settle the craft down and hope the ground is sufficiently flat. If you land at too great an angle you'll never be able to take off again. And, if things go amiss, there's no possible rescue."

Butler is equally concerned about the dangers the astronauts will face when it comes to leaving the Moon. He added: "They'll blast off and attempt to ride their exhaust flame up into lunar orbit. Will this all come to pass? Who knows? Going on the perfection of the programme to date, one can almost be optimistic. But one shouldn't be too surprised if Apollo 11 failed, if Armstrong and Aldrin were to die in a crash landing, if they were to be stranded. A boulder or a disguised crevasse could turn triumph to disaster, spelling a grim end."

CREW'S REFLEXES TESTED

The astronauts faced a tough series of tests today on what to do if the launch goes wrong.

Instructors threw in various emergency scenarios as the trio worked for eight hours in the cramped command module simulator.

Repeatedly rehearsing the first twelve minutes after blast-off, the crew faced snap decisions on whether to abort. George Hage, Mission Director, said: "We're concentrating on developing the reflexes when you've got to react quickly."

If a problem develops just before or after launch, the crew can trigger an escape rocket and the capsule would separate from the Saturn V and they'd parachute down. After that, the astronauts would have to use the command module engine to separate from the Saturn.

NASA has rescue crews at different locations on the ground and at sea in case of an abort. An Air Force 'beach gang' will be a few miles away equipped with three helicopters, two amphibious vehicles and a landing craft. The helicopters will each carry a surgeon and paramedic divers. Various ships and aircraft are on standby, with the Navy destroyer, USS *New*, the farthest away at 650 miles.

ROCKET UNDER THE MICROSCOPE

Hundreds of technicians today carried out painstaking pre-countdown checks on the rocket and spacecraft.

Tight security included strict monitoring of every engineer and piece of equipment allowed on to Launchpad 39A. Robert Abbott, head of Saturn V Quality Surveillance, said: "A man must have the right badges and the right reasons to get in. Strict records are maintained on all tools used by anyone working inside the rocket. Anything that goes inside – that doesn't fly – must come out again."

Chuck Henschell, Launchpad Test Supervisor, said: "It's like a surgeon counting all his scalpels to make sure he hasn't left any inside the patient. We sure don't want a spanner in the works – literally."

Engineers have been inspecting a maze of pumps, valves, fuel lines, wiring and electronic devices. Abbott added: "Any person who works on the Saturn knows his business and the inspector who monitors this work has to know the anatomy of the rocket."

SPLASHDOWN REHEARSAL

Three Navy divers pretended to be the Apollo 11 astronauts today in a full-scale splashdown rehearsal in the Pacific.

Captain Carl Seiberlich said the USS *Hornet*'s test sixty miles off Hawaii went without a hitch as the pretend astronauts and a replica spacecraft were plucked from the sea and airlifted on board.

The job is complicated by precautions against possible 'Moon bugs'. No-one can have direct contact with the astronauts, whose Mobile Quarantine Facility was loaded on to the *Hornet* yesterday. The ship has had to be stripped down to accommodate this and TV crews' equipment.

Seiberlich – a 48-year-old Californian who took charge of the *Hornet* in May – said: "This ship was in Vietnam for eight months and, when we came back, we had just one month in the yards to get the crew some leave time and work on her equipment. We looked at everything and asked 'Does this contribute to Apollo?' If the answer was 'No', then off it went."

For splashdown, the *Hornet* will carry eight helicopters, four radar reconnaissance planes and three transport aircraft. The reconnaissance planes have huge mushroom-shaped radar pods fitted to them to get a fix on the descending spacecraft. The *Hornet* has returned to Pearl Harbor, but leaves on Saturday in case an emergency recovery is needed soon after launch.

LAUNCHPAD PRANKS

The last man the astronauts will see before blast-off was keeping tight-lipped today about what surprises he might have in store for them.

Pad Close-out Crew Leader Guenther Wendt will help settle the astronauts into the spacecraft before sealing them inside.

German-born mechanical engineer Wendt, 45, has seen off nearly every American manned spaceflight and is loved by the astronauts for his attention to detail…and his practical jokes.

In March, when the Apollo 9 astronauts crawled into the command module, they found Wendt had left a roll of sticky tape and a bottle of glue with a note reading: "For places we missed."

That was the latest in a string of tit-for-tat jokes which date back to the early Mercury flights. For Gordon Cooper's first flight in 1963,

Pad Close-out Crew Leader Guenther Wendt (right) chats with Neil Armstrong in the White Room during last week's launch dress rehearsal. Wendt has a reputation for playing pranks on astronauts.

Wendt presented him with a ticket reading: 'Present this and 25 cents at Pad No.14 for the ride of your life.' When Cooper returned, he gave Wendt a quarter he had taken into Earth orbit and a plaque engraved with: 'The fare for the ride of my life.'

Wendt says the astronauts tease him about his German background – he's affectionately known as the 'Pad Führer' – but they know they can rely on him. He'll demand that his six-man Close-out Crew, including astronaut Fred Haise, arrive several hours early on launch day. He said: "They all have special functions and if one should have a car accident and not show then I'd have to pull in a back-up man."

Two hours before the astronauts arrive, the team will take an elevator to the White Room – the launch tower chamber attached to the command module – for final checks. Once the astronauts join them, Haise and a technician will get inside the cramped command module to help the astronauts settle in and hook up their equipment before crawling out under the seats. With fifty-five minutes to go, the Close-out Crew will leave. Wendt added: "I'll ask each crewman if everything's okay. There are no re-runs once we close the hatch."

QUAKE EXPERIMENTS

Scientists are hoping a $2 million Moon experiment could help predict earthquakes.

Rotational wobbles of the Earth and Moon are thought to be connected to seismic activity here and a lunar-ranging reflector will be left on the lunar surface to accurately monitor these wobbles.

The astronauts will set up the equipment, comprising one hundred cut-glass prisms, and laser beams will be fired at it from observatories on Earth. By measuring the time it takes for the light to bounce back, scientists should be able to measure the distance between the Earth and Moon to within a matter of feet. The two bodies are about 238,857 miles apart, but that can vary by as much as 10,000 miles.

Doctor Donald Wise, Lunar Exploration Office Chief Scientist, explained: "We know there's a relationship between earthquakes and the Earth's wobble. If we can detect a change in the wobble before an earthquake occurs, we might be able to establish an early-warning system."

Physicist Doctor Don Lind, one of the scientist-astronauts recruited by NASA for future missions, said: "The reflector will be an international facility. Anyone can bounce signals off it, so anyone can conduct their own experiments."

A second device – a lunar seismometer – will monitor surface disturbances in a bid to establish whether the Moon has a molten core, like the Earth. Wise said: "If we get fairly steady activity from one area, we'll know the Moon has quakes. Individual big bangs will indicate meteorite impacts." Don Breseke, Project Engineer for contractor Bendix Aerospace, said: "It'll be able to detect the impact of a meteorite weighing one gram – about the size of a pea – from a distance of 1km."

NATIVE CHIEF'S WARNING

A Native North American Indian chief today warned there will be "terrible sickness" on Earth if men land on the Moon.

Living as one with nature is the cornerstone of native culture and Joseph Logan Junior, Medicine-Maker and Hereditary Chief of the

Longhouse People of the Iroquois, believes the landing won't happen because it would disturb the balance of nature.

He said: "The Moon is our grandmother. She controls the growing of the crops, the hunting of deer, the tides, the seasons and men's minds. The Moon is sacred to our people. We predict man won't land on the Moon."

The 1,500 Longhouse People are among 9,000 natives living on the Six Nations reserve in Ontario, Canada.

The mission has also caused concern among elders of the Hopi people, who believe their descendants built the Moon using tanned buckskin under the instructions of Ee-Tana – 'Our Father' – to act as a shield to keep them safe on Earth.

Preston Keevama, a Hopi who works at the Los Alamos scientific laboratory in New Mexico, said: "Our old ones hope men who go to the Moon will treat it with dignity and honour, so that Ee-Tana will leave it there to guide the destiny of mankind on the rightful path."

ASTRO POSTMEN

Neil Armstrong and Buzz Aldrin are to become the first Moon postmen.

Postmaster General Winton Blount announced today that the astronauts will take an engraved master die to the Moon and use it to stamp an envelope. The envelope – franked with the postmark 'Moon Landing. USA. July 20, 1969' – will be brought back to Earth and the die will be used to produce a commemorative airmail stamp.

Production of the stamps – double the size of regular ones – will be delayed until August because the die will be in quarantine. Designed by Paul Calle, from Stamford, Connecticut, the stamp depicts an astronaut at the bottom of the lunar module ladder. An initial print run of 120 million is planned.

SPACE MONKEY DIES

Bonny the space monkey died today in a Hawaiian laboratory, twelve hours after he was recovered from the Pacific.

NASA was shocked when the animal suddenly passed away after the team treating him in intensive care initially reported he was in a stable condition.

Brad Evans, Public Information Officer, said: "It was sudden. He'd been in fair condition just before. We had no idea he was that close to death. He just suddenly fell off."

Bonny's Biosatellite 3 spacecraft was plucked from the water in a dramatic rescue operation near Kauai Island. A waiting aircraft, hoping to catch the capsule in mid-air, lost sight of it in heavy cloud. The capsule rode eight-foot waves for fifty-five minutes before being airlifted to Hickam Air Base. It took scientists nearly two hours to remove a maze of wires and life-support systems from the animal.

NASA doesn't believe the monkey was injured during splashdown. Doctor William Ross Adey, Project Principal Investigator, said: "It's a complete mystery why the monkey should have gone so soon, so fast. There was no specific evidence of heart or brain malfunction."

Bonny's death is a sobering reminder of the perils of spaceflight, but a NASA spokesman said: "The Biosatellite mission has no effect whatsoever on Apollo 11. We've had many manned flights of ten to fourteen days and there were no ill effects on the crew."

BAIKONUR BLOW

Frank Borman flew back to Moscow tonight with his hopes of visiting the main Soviet space centre seemingly dashed.

The astronaut spent the penultimate full day of his USSR tour in Siberia and was left disappointed when his flight there from the Black Sea region passed close to the secretive Baikonur facility without stopping off.

Borman and his family were warmly received in Novosibirsk, one of the country's most modern cities. Novosibirsk, which means 'New Siberia' and has a population of more than a million, is largely built on former marshland and is a symbol of the Soviets' goal to develop the region.

The Bormans met scientists in an elite village called Akademgorodok – 'Academic Town' – and the astronaut told them: "Science is science and, if you forget language, it's the same everywhere."

Asked today what Soviets think about the Moon shot, Borman said: "They wish us all success, just as we've wished them success in all their flights."

CHAPTER 9

WEDNESDAY, 9 JULY
Launch Minus Seven Days

"The flight simulators are as identical to the real thing as we can humanly make them. We throw in an added problem to show them a thing or two."

NASA Simulations Director Riley McCafferty

"Women are used to having the last word. Sometimes we even cry a little if it's the only way to get things done."

Apollo Radar Systems Expert Amy Spear

The lunar module simulator at the Manned Spacecraft Center in Houston. Technicians use computers to feed the astronauts various mission scenarios.

ROCKET LEAK SCARE

NASA tonight confirmed countdown will go ahead on time, despite a scare over a leak in the Saturn V.

During a tense day at Cape Kennedy, engineers discovered a fault in the system which pressurises liquid oxygen fuel used in the rocket's first stage.

With countdown due to start at 8pm tomorrow, NASA initially admitted: "We don't know where the leak is or how serious it is."

However, two technicians climbed inside the 64ft-high, 33ft-wide liquid oxygen tank to pinpoint and fix the leak. It turned out to be a minor issue and Doctor Kurt Debus, Kennedy Space Center Director, insisted tonight: "We can live with this sort of thing."

Today's unlikely heroes were two technicians from Boeing, contractor for the Saturn V first stage. Ira Ray Beeson and Walt Delle clambered into the tank and used rib-like panels on the interior walls to climb. They traced the leak to one of four helium bottles and were able to stop it by simply tightening a loose nut.

If launch were delayed, the only alternative dates in the near future are 18 and 21 July. The Moon has to be in a favourable position to attempt a landing and a delay would also mean switching landing sites to ensure the right light conditions on the surface. The astronauts are eager to stay on schedule because the Sea of Tranquility is the landing site for which they have prepared most thoroughly.

BATTLE OF WITS

An astronaut training chief explained today how his instructors have been waging a daily battle of wits with the Apollo 11 crew.

Neil Armstrong, Buzz Aldrin and Michael Collins spent today working in flight simulators and Riley McCafferty, Director of Simulations, revealed his instructors are under orders to show no mercy and throw unexpected scenarios at the astronauts.

Computers in the command and lunar module simulators can challenge the astronauts and Mission Control with as many as 1,700 different malfunctions. McCafferty said: "Our job is to study the men, see where they're weak and where they're strong and to see if the procedures for the mission are adequate. If we see a trend to an error on the part of an astronaut or a controller, we may throw in an added problem to show them a thing or two."

Each astronaut has spent hundreds of hours working in the simulators and McCaffery added: "They're as identical to the real thing as we can humanly make them."

Today, Armstrong and Aldrin's training included putting on their bulky backpacks inside the cramped lunar module. Collins, working in the command module simulator, rehearsed orbiting the Moon alone. The command module cockpit is an exact replica of the real thing but, on the outside, it's a 30ft-high structure weighing 40 tons. It houses TV cameras and electronics equipment used to duplicate the sights and sounds of spaceflight, plus five tons of mirrors and lenses. In each of the simulators, models of the Earth, space, Moon and lunar landscape are used to give realistic views depending on the scenario.

NEIL'S WARM, NOT COLD

Neil Armstrong has a big enough personality to live up to his history-making role, according to NASA's top doctor.

Some commentators have complained Armstrong has come across as cold and boring in interviews and press conferences, but Doctor Charles Berry, Medical Director, insisted: "When you know him well, he can be a very warm person and he has a great deal of humour, which seldom comes out in his superficial contacts with people. Almost everybody has seen Neil blush. He appears to be very sensitive. I've been told he gives the impression of being cold. I attribute this to bashfulness."

Apollo 7 astronaut Walt Cunningham believes people expect too much from astronauts. He said: "We're as human as anyone, but part of our discipline is to treat the most fabulous spectacles in the universe in the most matter-of-fact manner. If I were a poet or some other highly creative type, I might get so carried away I'd goof on the essentials."

NBC TV newsman Roy Neal, who counts Neil Armstrong as a friend after first meeting him when the astronaut was a test pilot ten years ago, expects the crew to go about the mission in a no-frills, business-like manner. He said: "They're all pure engineers and their personalities will be reflected in their language. They'll probably make it all sound like an everyday occurrence."

WOMEN FLY HIGH

A working mum told today how women are breaking down equality barriers in the male-dominated space programme.

Amy Spear has helped electronics firm RCA, a key NASA contractor, develop radar systems for when the command and lunar modules rendezvous and dock.

Spear admits she's had to overcome sexism from some men among the 2,000-plus staff working on the space programme at RCA's plant in Burlington, Massachusetts. The married mother-of-four, who has a Master's in electrical engineering, said: "I guess there are times I'm rough on some of the men, but perhaps it takes a woman to be so painstaking, to make absolutely sure that everything's the way it should be.

"I've had arguments and fights about the way things should be done. Usually I've won on engineering knowledge, through pure persistence. I'm responsible for all quality control. I stick my finger into everything here to make certain things are done properly. Women are used to having the last word. Sometimes we even cry a little if it's the only way to get things done."

WORLD'S MOST EXPENSIVE TAILOR

Neil Armstrong and Buzz Aldrin are watching their weight, under strict orders from their Moon 'tailor'.

When the pair step on to the lunar surface, they'll be wearing special suits costing $100,000 each and designed to protect them from the extreme conditions. As with the regular spacesuits the crew will wear during other phases of the flight, they have been individually tailored.

NASA's Charles Lutz, who can claim to be the world's most expensive tailor, explained: "Every suit is custom-made for the man and must fit exactly. We don't like them to put on weight, although that isn't usually a problem. We try to meet individual wishes. Some want long gloves, some want short ones. Some want their electrical harness attachment in one place, some in another. But we can't concern ourselves too much with personal whims of styling. The suit has to keep to the design."

The lunar suits have fifteen layers, starting with delicate nylon chiffon closest to the astronauts' skin and ending with a top layer made of Super Beta, a fibreglass fabric which is coated with Teflon. The suits – made by

Delaware-based ICL – must cope with temperatures varying from 320 to minus 250 degrees Fahrenheit. They must also withstand a continuous hail of micrometeorites, tiny rock and metal fragments which bombard the Moon at 45 miles per second.

Armstrong and Aldrin's lunar gear – suit, oxygen and air conditioning backpack, helmet, gloves and boots – will weigh 183lb compared to a regular six-layer spacesuit's 35lb. In near-zero gravity, however, the actual weight will be only a sixth of what it feels like on Earth.

The pair will wear water-cooled undergarments to help regulate body temperature and, to make their work on the surface easier, plaster casts were taken of their hands to ensure their gloves fit exactly. Their boots have tractor-tread soles to stop them from slipping. The suits have accordion-like joints to give the astronauts greater range of movement, with ICL drawing on its experience in making bras and girdles to achieve this.

Hundreds of different tests have been carried out on the equipment and even Doctor George Mueller, a NASA Associate Administrator, has tried on a Moon suit and backpack. Mueller said: "I wouldn't ask anyone to do anything I wouldn't do myself."

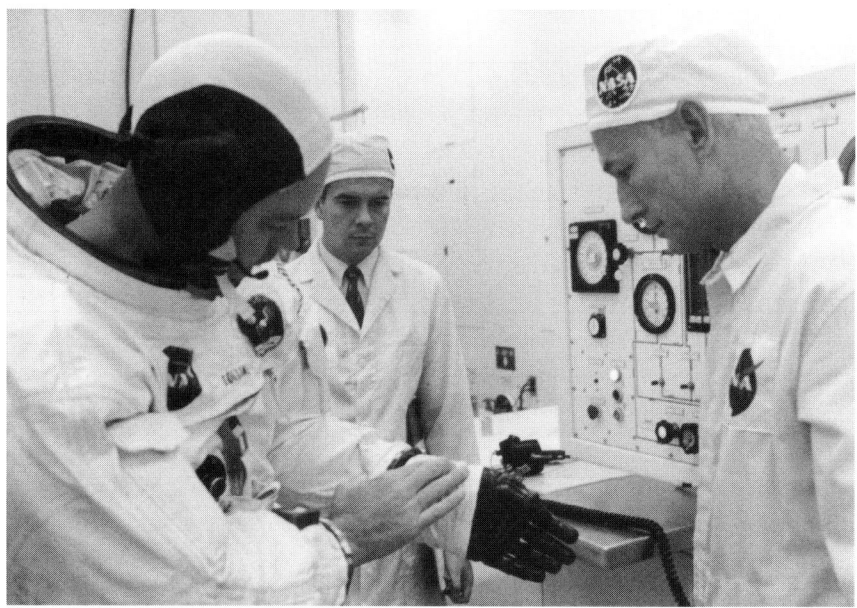

Michael Collins in discussion with technician Joe Schmitt as he tries on his spacesuit. The crew's gear is all individually tailored.

BORMAN'S KREMLIN MEETING

Soviet President Nikolai Podgorny met Frank Borman today and wished Apollo 11 success.

The astronaut was granted a forty-minute audience at the Kremlin on the final day of his USSR tour and described the meeting as "encouraging and beneficial when you think of space co-operation".

State news agency *Tass* said Borman told the President he felt they were all friends and Podgorny replied: "I think so, too, especially since our cosmonauts, who are friends of yours, are also heroes."

Tass said Podgorny had "wished success to the planned spaceflight of Apollo 11", adding: "There have been different periods in the relationship between the Soviet Union and the United States, but there have always been kindly feelings between the peoples."

In a lighter moment, when Podgorny offered Borman's wife, Susan, a strong Russian cigarette and lit it for her, the astronaut joked: "You're ruining my wife's habits!"

Alongside Leonid Brezhnev and Alexei Kosygin, Podgorny was part of the 'Troika' who ousted Nikita Khrushchev in 1964. It is thought his role as head of state is largely ceremonial, while Brezhnev and Kosygin wield the real power.

Borman, who earlier also met top space scientists, said in an interview published today by Soviet newspaper *Literaturnaya Gazeta:* "We should stop unnecessary duplication in planetary exploration. I'd like to believe that, in the not-distant future, there will be a time when scientific laboratories will be in orbit and scientists from different countries in the world will co-operate aboard these ships."

Borman's trip was arranged by the Soviet Society for Friendship. Deputy Chairman Viktor Gorshkov thanked the family for visiting and apologised for the busy itinerary. He joked: "Your trip probably has been almost as difficult as your flight to the Moon."

While events in Moscow today were cordial, there was a stark reminder of the superpowers' uneasy relationship as the US Navy confirmed it was monitoring a task force of Soviet warships spotted off the Florida coast.

The fleet, including two guided-missile destroyers, may be heading to Cuba for the 26 July celebrations of the revolution, but it's the closest the Soviets have ever operated to the east coast.

FURY OVER MONKEY'S DEATH

Animal lovers today slammed NASA over the ill-fated flight of Bonny the monkey.

Don Zylstra, NASA Public Affairs Officer, admitted: "We've had close to 1,000 letters and some nasty telephone calls. We're conscious of this concern. However, we're astonished that there seems to be more concern for the animal than for our astronauts."

Cameron McNeill, President of the Washington Humane Society, complained: "It's an inhuman thing and we abhor it. The human being is self-willed. The monkey had no say. He was trained for months, then strapped into the capsule and sent spinning around the globe."

Only Bonny's arms were left free during the flight and Christine Stevens, President of the Animal Welfare Institute, said: "Astronauts aren't immobilised. They have freedom of choice. They're mentally and emotionally equipped for such a venture."

Doctor William Ross Adey, Project Principal Investigator, who was in charge of an autopsy carried out today, said: "The preliminary findings indicated that the animal was, in every respect, healthy."

Small bruises were found on Bonny's liver and heart, but a NASA statement said: "The minor injuries to internal organs were probably due to re-entry forces. Death was probably due to the cooling of the animal and consequent heart failure."

'I OWN THE MOON'

A Chilean lawyer has written to Richard Nixon and NASA to remind them he *owns* the Moon.

Jenaro Gajardo Vera gained celebrity status in his homeland in 1954 when a Chilean court granted him ownership of the celestial body.

The law there states that anyone can stake a claim to a property which has no registered owner. He declared his claim and, when no-one challenged him within thirty days, a court granted him ownership.

He has now sent a telegram to NASA saying: "I wish you a happy arrival to my domain." He also wrote to Nixon asking for an official invite to the USA "on which occasion I'll take my title to share it happily with the champions of outer space".

WORLDWIDE SPACE CRAZE

Space fever is sweeping the world.

In Japan, a TV presenter plans to spend the duration of the flight wearing a spacesuit and eating space food.

Football-mad Colombia, where matches are played on Sundays, has taken the unprecedented step of calling off all fixtures on landing day.

Yugoslavia's biggest newspaper, *Vecernje Novosti*, is offering a cash prize to any reader who correctly predicts Neil Armstrong's first words on the Moon.

In England, people in London's West End will see mission bulletins flashed up on the screen of Europe's biggest electronic newscaster at the Swiss Centre, Leicester Square.

And in Melbourne, Australia, the National Cheese Club has sent a telegram to the astronauts saying: "If you find the Moon's made from green cheese we'd welcome a sample."

CHAPTER 10

THURSDAY, 10 JULY
Launch Minus Six Days

"Hell, this is it! It's almost as if I were a kid again looking forward to Christmas."
  Cape Kennedy Chamber of Commerce Official Charles Johnson

"I firmly believe we're going to have hotels in outer space."
                Hilton Hotels President Barron Hilton

Buzz Aldrin gets used to the feeling of working in weightlessness as he takes a flight on a jet transport plane.

COUNTDOWN UNDER WAY

Launch Director Rocco Petrone set the countdown clock running tonight and admitted his team is feeling extra pressure.

The former Army Colonel confirmed everything was on schedule for Wednesday's launch and that the enormity of the mission was beginning to hit home.

He said: "We've closed out anything of a worry nature, of a nature that would be no-go for the flight. The ground crew understands what's riding on this. I wouldn't be frank if I didn't say this is the pay-off. The risks are with us in any launch, so you'll always have some degree of apprehension, but the fact it's going to the Moon makes it probably mean more.

"We might not be able to land Apollo 11 for one reason or another, so we're preparing to repeat the attempt with Apollo 12, yet we earnestly hope and expect that this one will do the job."

The astronauts, who will give a final press conference on Monday via closed-circuit TV, spent most of today in the command module simulator practising the rocket burn which will fire them out of Earth orbit. Buzz Aldrin also went up in a jet transport plane to get used to the feeling of weightlessness.

COCOA BEACH BUZZING

Space tourists are already pouring into Florida in the rush to secure a grandstand view of the launch.

The community of around 10,000 in Cocoa Beach, which is just fifteen miles from Kennedy Space Center, expects a million visitors to descend this week. Roads into central Florida were today already jammed with cars, buses and motorhomes.

The sense of excitement among locals is obvious, but there's also concern over whether the area can cope and about what the future holds.

Launch week is set to bring around $5 million into the Brevard County economy and Charles Johnson, of Cape Kennedy Chamber of Commerce, enthused: "Hell, this is it! It's almost as if I were a kid again looking forward to Christmas."

Doctor Burton Podnos, a local psychiatrist and Director of the Brevard County Mental Health Center, said: "The excitement doesn't even have to be talked about; it's in the air. The kids are acting out the excitement and tensions of parents who are involved in the space programme. We've had calls from parents saying 'I don't know what to do with Johnny, he's uncontrollable this week'."

Officials responsible for managing traffic, safety and services are fretting. Phil Streit, County Safety Director, said: "We've covered all contingencies. We've planned as much as we can plan. Now we have to wait and see if it works."

Around 300,000 extra vehicles are expected in the area and Herbert Johnson, Brevard County Civil Defense Center Director, said: "That's the equivalent of 1,000 miles of automobiles parked bumper-to-bumper. Well, we only have 1,000 miles of highways here, so cars could physically cover every road in the county."

Visitors planning to watch the launch from boats have been warned by Lieutenant Donald Dean, Coast Guard chief, that an exclusion zone will be in force on local waterways and that anyone who infringes it will face a fine or even jail.

There are concerns over accommodation and food supplies. Hotels are fully booked as far away as Daytona, seventy-five miles to the north, and Orlando, sixty miles to the west. Maybell Wilkins, of the Cocoa Beach Chamber of Commerce, said: "We're getting phone calls from all over: Europe, South America. Just yesterday I called the Palm East Apartments and told the girl there that the President of the Bank of Nevada needed a room … she just laughed and said they couldn't even give movie producer Darryl Zanuck a room. Can you imagine?"

Rooms made available by around 300 local householders – some for free, others for up to $25 a night – are all taken. Some motels are setting up makeshift beds around their pools.

Walt Crosley, part-owner of George's Steak House, has ordered 3,000 extra steaks and said: "We feel there won't be enough food in the area. Trucks won't be able to get in. We've got to stock up now."

Space dominates Cocoa Beach, where the Moon Hut Café declares itself 'the home of the Moon burger', the Galaxy Lounge serves 'Lift-Off Martinis' and motels include The Polaris and The Satellite. But locals fear what will happen after the Moon shot. The population of Cocoa Beach has grown eighteen-fold since 1950, but the economic boom has been slowing since the early 1960s. The value of an acre of beachfront land has dropped from $6,500 to $4,500 in three years and 5,000 jobs are being cut at the space centre.

Bill Taylor, a local lawyer, said: "In the back of everyone's mind, they're concerned. People just don't know whether they're going to stay on here or not." Susie Johann, an assistant at Brevard County Library, admitted: "Suddenly, with the Moon shot happening, there's a sense that this is the last big event, that things are just blowing away."

Thousands of people – many of them African Americans from Mississippi, Georgia and the Carolinas – moved to Brevard to find employment. Now their jobs as labourers, maids and janitors are disappearing. Dorothy Sweetwine, of the Brevard Community Action Agency, said: "We're at a critical point. The county has no industry to hire the lower-level people who came here five years ago. Space is our only industry and it gave people a lot of work. What are they going to do now?"

There are bright spots for the local economy. The space centre has become one of the USA's most popular tourist attractions, with 1.9 million people taking the $2.50 bus tours TWA has been operating for NASA for three years. In January, direct flights start between Florida and England, with Lewis Bergman Maytag, President of National Airlines, predicting this will "open new opportunities throughout the Sunshine State". And a new theme park is being constructed in Orlando. The $600-million Walt Disney World, due to open in 1971, will be more than 50 times the size of Disneyland, California.

PUBLIC 'COULDN'T TAKE FAILURE'

Doctor Wernher Von Braun today admitted it will be a crushing blow to public morale if Apollo 11 fails.

The rocket genius told a press conference he is concerned that recent successes have made the nation complacent about the possibility of a space disaster.

He said: "No flight has ever been prepared with such attention to detail. We're confident, but there are a lot of pitfalls in a pioneering project. Chances have to be taken. There has been an unbelievable string of successes. If anything happens, the public may not be prepared for the shock."

Answering critics of space spending, Von Braun claimed the USA was "$24 billion richer, not $24 billion poorer", because space technology had created new opportunities for industry.

Asked if he'd like to fly into space himself, he said: "Yes, I'll be looking forward to the opportunity to thumb a ride. This business of spaceflight only being available to young heroes is probably on its way out. A space station will be inhabited by all sorts of people who will ride as passengers."

MOON COLONIES 'BY 1978'

Moon colonies will be up and running within a decade, a NASA expert predicted today.

Doctor Rodney Johnson, a NASA Advance Planner, believes there will be several 24-man camps established by 1978. He said: "A lunar colony will grow progressively from a few relatively simple shelters, laboratories, power supplies and life-support units to larger numbers of more complex ones."

He is confident new technology will soon be developed to recycle water, food and air taken to the Moon from the Earth. This could include purifying urine for use as water.

Doctor Paul Lowman, a NASA astro-geologist, insisted: "Our sister planet may indeed be a barren and inhospitable world in 1969, but so were the shores of New England in 1620. To the engineers of the twenty-first century, the Moon may be a rich and productive world."

Doctor Thomas Paine, NASA Administrator and a nuclear physicist, agreed: "We're opening a road to the planets; a road that man will take. This is something of great importance to humanity, indeed to life. The Moon is just as capable of supporting life – with modern technology – as North America was of supporting the Pilgrim Fathers."

Paine believes it will be a joint USA-Soviet effort. "Both sides realise that when we each have colonies on the Moon these will want to mutually support each other, at least in an emergency. I think we'll see increasingly that the exploration of space is done in the name of mankind, not of any individual nation."

THE LUNAR HILTON

The world's biggest hotel company plans to expand its empire to the Moon.

Hilton and American charter airline TIA have produced a glossy brochure inviting people to register for a Moon trip "sometime after 1973". Available in the lobby of the Cape Kennedy Hilton, it tells potential space tourists: "Enjoy your stay on the Moon, one of the most relaxing resorts in the universe. Its unique attributes include a smog-free atmosphere, no rain or snow, no breath of wind and profound silence."

Barron Hilton, the 41-year-old President of the hotel chain, predicted today: "I firmly believe we're going to have hotels in outer space, perhaps even soon enough for me to officiate at the formal opening of the first." And Darlene Robertson of TIA said: "The cost per person would be up to $25,000, but imagine flying to the Moon!"

DIVORCE RATE ROCKETS

A Kennedy Space Center engineer told today how the space race has become one of Florida's biggest marriage wreckers.

Thousands of support staff have been putting in long hours to help America beat JFK's lunar landing deadline and Chuck Henschell, Launchpad Test Supervisor, says many of his men have paid a heavy price.

Henschell, whose 500-strong team are busy making final checks on the spacecraft, said: "We're confident we'll get the Moon shot off on time. Every man here feels part of it. That this is what he's been working towards for years.

"Of course, their wives don't always see it that way. The women don't see why their men should spend so much time working here, while the men think their home life should be run in the same rigid way as the countdown. So, we do have a very high divorce rate, I'm afraid."

ROUND EARTH 'LIARS'

NASA has been accused of faking pictures which appear to prove the Earth is round.

Samuel Shenton formed the International Flat Earth Research Society in 1956 and says his members remain convinced the Earth is a stationary, dish-shaped planet.

Speaking at his home in Dover, England, the retired sign writer said: "The United States is practising a great deception. They are simulating the Earth as round and this is a great delusion for the world. These Americans and Russians are so damned cunning. For some reason or other, they obviously want us to think the world is round. Some of the pictures have been blatantly doctored. Studio shots, probably."

The Society believes the Earth is much closer to the Sun and Moon than is now widely accepted and that they are both small spherical objects of only about thirty-two miles in diameter. Shenton explained: "According to my calculations, the Sun is only 3,240 miles from Earth. Can you imagine what sort of summers we would have if it were 93 million miles away, as these scientists would have us believe?"

Shenton, 68, says the behaviour of balloons proves the theory he has believed in for fifty years. "If there is no wind, they stay hovering exactly where they are. And why? Because the Earth is stationary."

He admits the space programme has seen membership dwindle. "We've lost a lot of members because of this absurd Apollo. It's the same with every new space probe. We're down to fewer than a hundred members now, but if their opinion can be swayed by such flimsy evidence, then they're no use to the Society. I just can't get the message across to young people these days. Science teachers have deluded them from the cradle."

The Flat Earth Society claims that images, such as this one taken by Apollo 8, showing the Earth is round are fakes.

LUNAR CROONER

Veteran jazz legend Duke Ellington will sing in public for the first time in his long career when he gives a special TV performance to mark the Moon landing.

ABC has enlisted the 70-year-old composer, performer and conductor to write an original piece and he will perform *Moon Maiden* live on a lunar landscape set. For the first time in a career which has seen him write around 5,000 songs, Ellington will feature on vocals. He said: "It's a one-shot thing. This is the end of my singing career!"

The lyrics feature a mysterious Moon Maiden and Ellington added: "There's got to be a girl out there for a man to be singing about."

BORMAN BACK IN USA

Frank Borman was back in the USA tonight after a Soviet Union tour he described as a "complete success and a personal pleasure".

During a hectic day, the astronaut and his family flew from Moscow to England and then on to New York. Speaking in New York, at one of three press conferences held in three countries and in three different time zones, he said: "I believe we've laid the groundwork for detailed discussions on the question of more co-operation in space. Outer space is one area where we ought to be able to work together."

The family were seen off at Moscow's Sheremetyevo Airport by a delegation including three cosmonauts. Borman's wife, Susan, had a tearful farewell with their wives and said: "I don't think this family has enjoyed ten days in our lives so much as this." Borman urged the cosmonauts to visit the USA, joking: "I can introduce you to the wonders of tequila…that's Mexican vodka."

CHAPTER 11

FRIDAY, 11 JULY
Launch Minus Five Days

"If it were up to me, I'd aim for Mars instead of the Moon. Mars is more interesting from every point of view."
NASA Space Science Laboratory Director Doctor Ernst Stuhlinger

"By all means, progress in space, but not at the expense of spending money here on Earth on problems of poverty and ignorance."
Explorer Sir Edmund Hillary

Ethiopian President Haile Selassie (hand raised) asks a question of Launch Director Rocco Petrone (standing far right) during his visit to the Kennedy Space Center today. Astronaut Gordon Cooper is on the President's right and on his immediate left is KSC Director Doctor Kurt Debus.

CREW PASS MEDICAL

The astronauts passed their final pre-flight medical today, but have been ordered to get more rest.

After a four-hour examination, NASA Medical Director Doctor Charles Berry declared Neil Armstrong, Buzz Aldrin and Michael Collins ready to fly. However, he said they would be cutting their workload and taking a day off on Sunday.

When Apollo 8 flew to the Moon at Christmas it accelerated the whole programme. That mission was originally intended to be an Earth-orbit flight, but when the lunar module wasn't ready, NASA gambled on going to the Moon instead. That effectively brought forward the first lunar landing attempt, with Armstrong, Aldrin and Collins working overtime to prepare.

Berry said: "Our one concern was fatigue. We don't want to launch a tired crew. It's a difficult schedule these guys have gone through, but they've done better than we – and they – thought they would. Their physical state looks good. Today's medical indicates that the Apollo 11 crew is ready for flight. They really looked good."

The astronauts gave blood samples, had an electro-cardiogram, hearing and eye tests, chest X-rays and a urological examination. Skin samples were taken to compare with post-flight samples.

Later, the astronauts took a break from working in the simulators to speak to Ethiopian President Haile Selassie, who was visiting the space centre. He wanted to meet them, but had to wish them luck by phone because of quarantine restrictions.

Rocco Petrone, Launch Director, confirmed tonight that countdown was progressing well, but said: "You don't assume anything's going to work. You test it. It takes just one component to spell disaster. When you have that, boy you don't ever get confident to the degree of saying 'I can turn my back on this thing'."

The countdown lasts ninety-three hours, but this allows forty spare hours for stops known as 'hold time' during which engineers can fix any unexpected problems and take rest periods.

SOVIET 'YES' FOR LAUNCH

The Soviet Ambassador to the USA is to attend the launch.

NASA revealed today that Anatoly Dobrynin has accepted an invitation to watch, making him the first Soviet official to visit

Cape Kennedy. The USSR has repeatedly turned down previous invitations to visit.

The agency hopes it may now be invited to the Soviets' Baikonur base. NASA spokesman Julian Scheer said: "We certainly wouldn't ask, but this enhances the chances that an American will attend one of their launches."

Doctor Thomas Paine, NASA Administrator, believes the Soviet space programme has stalled for some reason, but said: "I think that once their problems are overcome, we'll see a very vigorous Soviet programme on the Moon. This will perhaps be the time they may be amenable to do something with us, after both of us have made a landing."

HUNGER MARCH

A civil rights group today revealed plans for a pre-launch protest march at Cape Kennedy.

The Southern Christian Leadership Conference (SCLC) says hundreds of what it described as "hungry people" from five southern states will be transported by bus to protest over billions being spent on space rather than social problems. Mules and wagons will be included in the march as symbols of how many people still live in extreme poverty.

Reverend Hosea Williams, SCLC Poor People's Campaign Director, said: "We're not against things like the space shot, but there has been a miscalculation in priorities."

All police leave has been cancelled as a precaution. Leigh Wilson, Brevard County Sheriff, said: "These people have a right to demonstrate, as long as they don't break the law or tie up the traffic."

EVEREST LEGEND SLAMS SPENDING

Legendary explorer Sir Edmund Hillary today said he wishes he was the first man on the Moon, but criticised the USA for spending billions on getting there.

The New Zealander, who climbed Mount Everest in 1953, admits his adventurous nature makes him jealous of Neil Armstrong and Buzz Aldrin.

But he insisted: "By all means, progress in space, but not at the expense of spending money here on Earth on problems of poverty and ignorance. I'm one of those who feel the speed at which spaceflight has progressed is the result of politics; competition between the United States and Russia."

Hillary, who will be fifty on the day of the landing, said he would have loved to have been the first man on the Moon "as most men who have the spirit of adventure would". He added: "It'll be something like stepping on the summit of Everest. You can't afford to be swept away by emotion. The important thing is getting down again. My first reaction was of relief that I'd reached the top and that I could go down again. And then there was surprise that it was me there, where so many others wanted to be."

JOHNSON INVITE BLUNDER

Richard Nixon today personally invited former President Lyndon Johnson to the launch after an embarrassing blunder by NASA and White House officials.

Thousands of VIP guests have been asked, including hundreds of US and foreign politicians, but Nixon's predecessor wasn't on the list, despite the key role he has played in the space programme. When Nixon realised, he phoned Johnson with an invitation which was instantly accepted.

It appears NASA and White House officials believed it was the other's responsibility to invite him. The blunder only emerged when reporters began to ask whether Johnson would be there. Johnson's office had said last week that he would consider an invitation "if one were offered".

In 1958, as a Senator, he took the lead in writing the National Space Act, which created NASA. As Vice President, he convinced John F. Kennedy to make the Moon landing a national goal and, once he was President, continued to champion space exploration. He stepped down before last year's election, with Republican Nixon defeating Democrat Johnson's Vice President, Hubert Humphrey.

MOON CAR ON STARTING GRID

NASA has launched a new space challenge – the race to design the first 'Moon car'.

For future landings, astronauts will need a surface vehicle to explore further afield and NASA is inviting bids for a $19-million contract to design and build a two-man Lunar Roving Vehicle which would be carried to the Moon on the lunar module.

Doctor John Naugle, Associate Administrator for Space Science, said: "An exploration programme can be more effective if it includes the

A 1967 artist's impression of how the lunar roving vehicle might look based on a design proposed by the Bendix Corporation, an American engineering and manufacturing company.

ability to traverse long distances, equivalent to the expedition through the American West at the start of the nineteenth century. It will also require the capability to penetrate into more rugged and dangerous regions."

Other more radical alternatives were rejected for cost and safety reasons. One was a Flying Lunar Excursion Platform proposed by Langley Research Center. It would have skimmed across the surface like a magic carpet, but would have been more expensive and even a small tumble from the platform could have caused a disastrous tear in an astronaut's spacesuit.

1,000/1 WINNER

A bookmaker is bracing itself to pay out £10,000 to an English punter who wagered £10 on a lunar landing happening before January 1971.

In 1961, John F. Kennedy's goal seemed fanciful, but science fiction fan David Threlfall monitored developments and, in 1964, asked bookmaker William Hill what odds it would offer. The bookie – confident it wouldn't happen and eager to entice other punters – quoted him odds of 1,000 /1.

Now the 26-year-old personnel officer from Preston, Lancashire, is set to win big on his £10 stake on 'a man, woman or child from any nation on Earth being on the Moon or any other planet, star or heavenly body of comparable distance from the Earth'.

A William Hill spokesman said: "We reckoned the true odds were more like 3/1 against, but to have offered that would have been absurd; we wouldn't have got any more bets. We're ready to pay. In fact, we're about to draw up the cheque. We're going to take a bad loss. The Americans just got going ahead by leaps and bounds and were too fast for us."

MAN ON MARS IN 1986?

America could land men on Mars in 1986, according to NASA's top rocket scientists.

Doctor Ernst Stuhlinger today unveiled his vision for a mission featuring an atomic-powered electric rocket capable of flying at 112,000mph, nearly five times the record 24,791mph achieved by Apollo 10.

Stuhlinger, Space Science Laboratory Director at Marshall Space Flight Center, told academic journal *Science*: "If it were up to me, I'd aim for Mars instead of the Moon. Mars is more interesting from every point of view."

He says electric propulsion would be needed to overcome fuel and cost issues, as well as to achieve the speeds necessary to make the mission duration viable. His article includes sketches of a huge dumbbell-shaped craft he says could make the round trip in nineteen months if attempted in 1986 when Mars is at its closest to Earth (33million miles away). Up to four astronauts would orbit for fifty days, making trips to and from the surface.

Stuhlinger is one of the Germans who came to the USA with Doctor Wernher Von Braun. Another of them, Kennedy Space Center Director, Doctor Kurt Debus, insisted today that a Mars landing is "an unavoidable future event".

It remains to be seen whether funds for such a venture will be made available, however. Since taking office in January, Richard Nixon has slashed around $50 million off the latest $3 billion-plus funding package Lyndon Johnson had proposed for NASA. The new President is also prioritising a controversial and costly plan to set up a dozen anti-ballistic missile sites in the USA to protect the nation from attack.

FAKE ROCK FEARS

Space chiefs have warned people not to be fooled by conmen claiming they have Moon rock for sale.

All genuine samples will be government property and no requests to buy samples will be granted. However, NASA fears a black market will spring up within days of splashdown. A spokesman said: "We do expect unscrupulous promoters and nuts will claim to have lunar samples for sale."

LUNAR DWARVES WARNING

Followers of a religion which claims dwarves live on the Moon have been told to prepare themselves for the possibility the prediction may not come true.

The Church of the New Jerusalem bases its teachings not only on the New and Old Testaments, but on the works of eighteenth-century Swedish theologian, philosopher and scientist Emanuel Swedenborg. He believed the Moon was populated by dwarves with 'voices like thunder'.

But, writing in his parish newsletter, an English minister says some of Britain's 4,000 followers of the Church of the New Jerusalem may be in for a let-down. Reverend Ian Johnson, from Birmingham, wrote: "The landing may possibly shake the faith of a very small minority in the authority of our prophet's writings. Others are likely to say that finding no dwarves will prove nothing, as the claim is that they were there in the eighteenth century. They might since have moved to another planet. I must confess to being almost indifferent about it."

CHAPTER 12

SATURDAY, 12 JULY
Launch Minus Four Days

"I know of no more effective way of spending $24 billion. I think, beyond a doubt, that the money that has been spent on space wouldn't make a significant difference in social welfare."

NASA Associate Administrator Doctor George Mueller

"It will get bastardly hot, but if you don't keep your cool you're not worth a damn."

Launchpad Rescue Team Leader Al Wozniak

Michael Collins (left) with Deke Slayton at Patrick Air Force Base after sharpening his reflexes by flying a T38 jet.

ASTRONAUTS EASE UP

Doctors' orders took priority today as the astronauts eased up on their workload.

On the advice of Medical Director Doctor Charles Berry, after yesterday's pre-flight examination, the crew did a reduced flight simulator session this morning.

Later, Buzz Aldrin stayed in his quarters to study flight plans, while Michael Collins sharpened his reflexes with a flight in a T38 jet and Neil Armstrong simulated the lunar landing approach by hovering close to the ground in a helicopter.

As launch engineers continued to prepare the Saturn V and spacecraft, George Low, Apollo Programme Manager, said memories of the Apollo 1 tragedy, which saw NASA implement a raft of changes in hardware and procedures after a damning board of review report, are a constant reminder that complacency can't be tolerated.

Low said: "Manned spaceflight is unforgiving. We once made a serious mistake and it took the lives of three astronauts. We've made every conceivable effort to avoid similar mistakes. Attention to detail, no matter how small, is mandatory."

George Hage, Mission Director, said: "The reliability demonstrated on the Apollo flights to date has far exceeded most people's wildest guesses. We're blessed with some superb hardware."

Exhaustive tests ensure the highest standards of workmanship and equipment reliability. The lunar module, for example, has been tested in altitude chambers and subjected to extreme heat and cold.

Doctor George Mueller, who oversees the management of the manned space programme, has personally tried the simulators, making a lunar landing and docking the spacecraft. He said: "We're not in the business of doing terribly risky things. All the crews – and their wives – have a high degree of confidence in the equipment."

SOVIET LAUNCH U-TURN

The Soviet Union will *not* be represented at launch after a shock U-turn.

NASA chiefs were delighted when Ambassador Anatoly Dobrynin accepted an invitation to attend, but today the Soviet Embassy said he was out of the country and wouldn't be back in time. No one will attend in his place.

The news came on the same day astronaut Frank Borman told a Cape Kennedy press conference that his visit to the USSR had raised his hopes of improved relations between the superpowers.

He said: "From the people on the subways to their President, all I heard was that they're wishing success for Apollo 11. I got the impression they're interested in the landing, not only as a triumph for America, but for all mankind. They'd like to be first, but they'll cheer the people who do it first. If the Russians landed on the Moon tomorrow, I'd be the first to congratulate them."

NIXON DINNER ROW

NASA's medical chief has been slammed over the cancellation of Richard Nixon's pre-flight dinner with the Moon crew.

Frank Borman hit out today after Doctor Charles Berry's public warnings that Nixon might pass on germs led to the engagement being scrapped.

The Apollo 8 hero said: "I delivered the invitation to the President from Neil Armstrong and the crew. The controversy occurred while I was in Russia and I was dismayed. The reaction was totally ridiculous. Medical advice should be made in private and not publicly. Now, how can you get the decision reversed? If anyone sneezes on the Moon, they'd blame the President."

Asked if the cancellation was "damned stupid", he responded: "You said 'damned stupid'. I'd use stronger words. A person who would isolate the astronauts from the President doesn't understand pilot psychology. It would've given them a lot of encouragement."

Berry issued a statement saying: "Everything that can be said on the subject has been said. I see no advantage to further discussion. I think we should get on with our work."

PARTY CAN WAIT

Celebrations will be put on hold in Neil Armstrong's home town until he returns safely to Earth.

The astronaut is the pride of Wapakoneta, a small Ohio town with 7,500 residents, and banners in the town centre proclaim 'God speed, Neil'. A parade was held in his honour after his 1966 Gemini 8 flight and more celebrations are planned for Apollo 11.

But Charles Brading, Homecoming Committee Chairman, said today: "There will be quiet concern until Neil steps from the spacecraft. I don't think wild demonstrations are in good taste until splashdown. The excitement will be on hold until everything is all right." Brading – who runs a drugstore where Armstrong worked part-time as a teenager – added: "It's the biggest thing ever to happen to this town. There's a feeling of pride."

Armstrong's parents, Stephen and Viola, visited Florida for the Gemini 8 launch, but this time they'll watch on TV at their home in Neil Armstrong Drive. Father and son did attend the Apollo 10 launch and Stephen said: "I was reassured by that. I'm confident Neil's mission has been as carefully planned as possible."

RESCUE TEAM STAND BY

A rescue team are ready to put their lives on the line to save the astronauts if the launch goes wrong.

While VIP guests watch the lift-off from five miles away, fourteen emergency rescue specialists will be on standby a third of a mile from the launchpad. They will take up position four hours before launch, sweltering inside three armoured personnel carriers in their heavy fire-fighting suits and helmets. Team Leader Al Wozniak, whose men gave a rescue demonstration to the media today, said: "It will get bastardly hot, but if you don't keep your cool you're not worth a damn."

The team – who have been rehearsing up to fifty drills a day for weeks – have a wealth of experience in dealing with emergencies and are prepared to risk their own lives. Recalling how he once dashed into the burning wreck of a crashed B52 bomber to drag a crewman to safety, former Air Force man Ed Bidault said simply: "I was just doing my job."

The team, including three specialist medics, were put together by security firm Wackenhut and are prepared for various evacuation scenarios. These range from simply picking up the Pad Crew and astronauts from the foot of the launchpad to climbing the stairs to the top of the 363-ft stack and dragging the astronauts down again. Possible escape routes include a zip-wire at the top of the tower and a chute leading to a reinforced underground bunker.

Wozniak said: "The control room tell us what rescue mode we should use, but I'm the guy on the spot and, in the end, I'm in charge of the whole nine yards. If I think there isn't time, for example, to use the elevator, riding the slide-wire down is a lot faster. Or maybe things look real bad and there's no time for anything but jumping into the chute and scooting into the blast-proof room."

HOUSTON, WEAVE A PROBLEM

A needlework expert has revealed how she answered a NASA SOS by using old-fashioned weaving skills to mend a piece of hi-tech lunar equipment.

Space agency technicians found it impossible to repair an expensive 10ft-tall mesh antenna the astronauts may have to erect on the lunar surface to transmit TV signals.

In desperation, NASA contacted Pennsylvanian firm Blue Ridge Winkler Textiles and Arlene Van Atta saved the day with the art of 'invisible mending', a sophisticated process dating back two centuries. It involves using individual threads to make a near-perfect repair.

The umbrella-shaped antenna which Neil Armstrong and Buzz Aldrin may have to deploy on the lunar surface.

The umbrella-shaped antennae have reflective mesh made of gold-plated thread thinner than human hair and this is prone to snagging like a pair of tights. Van Atta said: "When I was told these were being taken to the Moon, I thought it was fantasy! The mesh is almost like a cobweb and I had to work with a lighted magnifying glass to thread it in the needle. Working on them was hard on the eyes and, one time, I had to stand on two boxes just to reach the damaged area.

"Apparently what I did turned out okay because they continued to call me. They made twelve antennae and I've worked on all of them. I've saved them a pretty nice bundle. If I hadn't been able to fix them, they'd have had to build new ones."

The 'Lunar Umbrella' will be used if the signal from the lunar module's in-built antenna is too weak. Van Atta joked: "If they poke a hole in it up there on the Moon, I'll go up and fix it!"

DIVER IS OVER THE MOON

A young navy diver spoke today of his excitement at being chosen as the first man to see the astronauts when they return to Earth.

John Wolfram, 20, from Atkinson, Wisconsin, will be the first of the USS *Hornet* frogmen to reach the spacecraft after splashdown and will peer inside to check on the crew. He said: "It's really great! I'll be the first man on Earth to see them; the first in the entire world!"

The recovery ship is back at sea, heading into position in case it is needed for an emergency recovery soon after launch. Captain Carl Seiberlich said: "The crew are really charged up. They really want to get out there and do this job."

NASA CHIEF'S WELFARE CLAIM

A top space agency boss today claimed the billions pumped into the Moon shot wouldn't have solved social problems.

Opponents of space spending claim the money should have been used for welfare, housing and other social programmes.

But Doctor George Mueller, Associate Administrator for Manned Space Flight, insisted: "I think, beyond a doubt, that the money spent on space wouldn't make a significant difference in welfare and yet it will make, in the years ahead, a significant difference in the productivity of the nation.

"I know of no more effective way of spending $24 billion in advancing the basic productivity and advances in technology that are essential for creating the national wealth that will, in turn, make it possible for programmes in areas including the cure of cancer.

"Our basic economy depends on the creation of new technology and new science. Without it, we'll lose our competitive position among the nations of the world. We've trained 300,000 people in a whole new way of doing business. Right now, there are about 150,000 people trained in the space programme who are working elsewhere in industry."

Mueller says Sputnik was a wake-up call which revolutionised the education system and that the space programme has benefitted weather forecasting, navigation, communications, geology, map-making, agriculture, medicine, biology, astronomy, photography and electronics.

The Nixon administration has already trimmed the space budget, but Mueller warned against opening the door for the USSR to retake the lead in space. He said: "It's not a fear, it's a fact. We've been going down in funding by ten per cent per annum for the last three years, but to the outside world we've been forging ahead. If this continues, we'll inevitably be out of the manned spacecraft business.

"I think the next generation isn't only going to accept but expect the continuation of space exploration. I don't think anyone who saw the pictures, through the eyes of the Apollo 8 and 10 astronauts, of how small the world is in this vast ocean of space, can ever again feel quite the same way about the purpose of conflicts on the Earth. We're together on a very small planet and we had best find ways of working together."

SCREEN TEST

TV anchormen face their own space challenge – staying on screen for thirty hours without a break.

America's big three TV networks have splashed millions on their coverage, including building elaborate studio sets with full-sized replica spacecraft and deploying dozens of cameras at multiple locations.

ABC, NBC and CBS all plan marathon continuous broadcasts for the landing, Moonwalk and lunar blast-off. Jules Bergman, News Science Editor for ABC, says he is determined to stay on air throughout. Bergman, 39, vowed: "Management has made provisions for me to be relieved for about five hours, but I intend to stay awake and on air. I don't want to miss one piece of communications or data."

Veteran CBS newsman Walter Cronkite, who has anchored their space coverage since the first manned flight, also intends to stay on air. Cronkite, 52, said: "It'll depend on how the adrenalin is pumping, but I don't want to miss anything."

Throughout the mission, the three networks – plus America's four main radio broadcasters – will pool pictures and sound for use worldwide. As part of this arrangement, a 31-man ABC team, headed by Producer Ron Ogle and armed with three van-loads of equipment, are on board the USS *Hornet* recovery ship.

Ogle said: "The difficulty compared to past splashdowns is that we'll have to telecast the astronauts in glimpses, through windows, while they come aboard. There'll be no flight-deck ceremonies to welcome them home. The Mobile Quarantine Facility is 14ft from where the helicopter will land and, when they take that short walk, we'll get what may be our only good look at them."

The TV networks don't expect to recoup all the money they're spending. CBS has spent $2.5 million, NBC $2 million and ABC $1 million. They will get some back from syndication to foreign broadcasters and shows will be sponsored, which wouldn't be appropriate for some live news events. CBS lost $1.25 million when it ran twenty hours of continuous unsponsored coverage of the John F. Kennedy assassination.

More cameras will be used than for any previous event, with news teams deployed at locations including Cape Kennedy, Mission Control, the astronauts' homes and the splashdown site. CBS, for example, is using more than 100 cameras.

Life-size spacecraft replicas built by NASA contractors have cost both NBC and CBS $500,000, with ABC spending $250,000. CBS has built a fake Moonscape on a third of an acre at its New York HQ and ABC's command module replica has been fixed to a hydraulic lift so that it can imitate the real thing's angles of flight.

CHAPTER 13

SUNDAY, 13 JULY
Launch Minus Three Days

"This is going to be, I'm sure, one of the most traumatic and exciting events of American space history, if not American history...period."
   NASA Director of Flight Control Operations Doctor Chris Kraft

"The irony is so apparent here. We're spending all this money to go to the Moon and here, right here in Brevard County, I treat malnourished children with prominent ribs and pot bellies."
   Brevard County GP Doctor Henry Jenkins

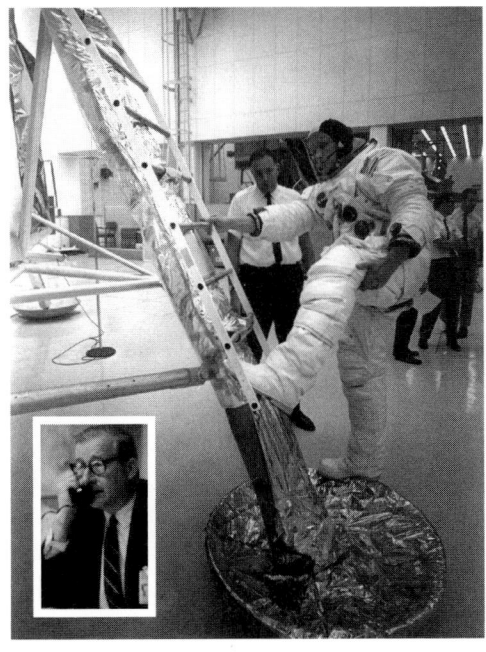

Neil Armstrong rehearses climbing the ladder of the lunar module in training this week. He hopes to do it for real next week, but Doctor Thomas Paine (inset) has stressed NASA won't hesitate to call off the landing if Apollo 11 runs into problems.

SOVIET MOON PROBE SHOCK

American space bosses were left stunned today after the USSR launched its own Moon mission.

The Soviets confirmed this morning that they launched an unmanned probe called Luna 15. *Tass* gave only sketchy details, saying the probe was already 41,000 miles from the Earth and that "the aim of the flight is to conduct further scientific exploration of the Moon and space near the Moon".

Frank Borman speculated: "I'd guess the probe is an attempt to land an automatic sampler and return some of the Moon's surface to Earth. It'll be a great feat if they can do it, but an unmanned machine certainly won't take the edge off Apollo 11. I don't think you could really take away the drama of a man walking on the Moon if you're trying to upstage somebody."

NASA issued a statement saying it wished the Soviet mission "every success", but Donald Stullken, Apollo 11 Recovery Team Leader, said: "A lot of people are going to be unhappy about this."

Rocco Petrone, Launch Director, claimed: "The Russians started the race in 1957 and today find themselves not fully in the race. It shows they're quite desperate."

Petrone, meanwhile, said Apollo 11 preparations were still ahead of schedule, allowing a sixteen-hour hold in the countdown to give engineers a break ahead of the crucial final forty-eight hours. He said: "We're in very fine shape, but we've got some big steps ahead of us."

DAD VISITS ALDRIN

Buzz Aldrin enjoyed some private time with his father today as the astronauts took it easy.

Aldrin, Neil Armstrong and Michael Collins spent most of the day relaxing in their quarters after poor weather put paid to their plans for a trip to the beach. They also worked out in the gymnasium.

Aldrin was visited by his father, a distinguished aviator who still holds a pilot's licence at 73 and whose advice the astronaut is known to value highly. Edwin Aldrin Senior was a close associate of legendary aviator Orville Wright and a friend of rocket pioneer Doctor Robert Goddard.

JFK DEADLINE 'IRRELEVANT'

Space agency chiefs won't hesitate to abort Apollo 11, regardless of John F. Kennedy's pre-1970 Moon landing deadline.

NASA is on course to hit JFK's target with five months to spare, but officials say no unnecessary risks will be taken to achieve his goal.

Doctor Thomas Paine, NASA Administrator, pledged: "We have no false pride about specific dates. We won't hesitate to postpone if we feel we're not ready and, once the voyage has begun, we have no pride that would make us hesitate to bring home the crew immediately if we encountered problems."

Doctor Chris Kraft, Director of Flight Operations, said: "This is going to be, I'm sure, one of the most traumatic and exciting events of American space history, if not American history...period. I'm worried about how well they're going to land, but that's up to the crew once they get to about 500ft. It's all in Neil Armstrong's hands."

Appearing on CBS show *Meet The Press* tonight, Apollo 9 astronaut Jim McDivitt said the landing and lunar take-off were the biggest concerns. Asked if he "thought the risks were worth it", he said: "We *know* it's worth it. You feel it in your heart."

LIFE GOES ON FOR WIVES

The astronauts' wives were today doing their best to get on with their daily business, despite a media siege outside their Houston homes.

Joan Aldrin attended Sunday service at the Webster Presbyterian Church, where her husband is a lay preacher, and Jan Armstrong ventured out to a local grocery store. NASA said Pat Collins was currently out of town.

In the latest issue of *Life* magazine, Joan gave an insight into how she plans to spend her time during the mission. She said: "I'm going to have an orgy of domesticity. I'm going to have all the rugs cleaned, paint the walls and wash the windows. Anything to keep busy.

"This time I'd prefer not to see the launch. I want to stay at home and live rather in a cocoon. Most of the time I have a marvellous faculty for putting out of my mind what I don't want to face. But some days it doesn't work. Then it's like walking on eggs, pretending normalcy. Normalcy tinged with hysteria!

"Just before the crew was announced I had one of those days. I wished Buzz were a carpenter, a truck driver, a scientist – anything but what he is.

Now I'm resigned to it. He wants to do this and I want him to do what he wants to do. I want him to do what he wants, but I don't want him to ..."

Of the impact on their lives, she added: "I really do believe that Buzz will still be the same person."

OCEAN 'BUGS' WARNING

Two scientists have urged Richard Nixon to order a change in the quarantine plans.

Doctor Leroy Augenstein and Doctor Barnett Rosenberg, biophysicists at Michigan State University, have written to the President claiming that opening the spacecraft hatch while it is floating in the Pacific will expose the planet to unnecessary risks.

Warning that the ocean could be a fertile breeding ground for lunar organisms, they wrote: "The admittedly small probability of life on the Moon could nevertheless lead to a major catastrophe for mankind if some lethal contamination occurred for which we have no natural defences."

One man who isn't worried is Doctor John Hirasaki, a Texan scientist who has agreed to go into quarantine with the astronauts. As Mobile Quarantine Facility Project Engineer, he helped design the crew's temporary home. Hirasaki, 28, said: "If I calculated it a high risk, I wouldn't be here."

KIDS 'STARVING' IN BREVARD

A doctor today claimed children living on the doorstep of the launch site are starving.

Critics of the space programme say it's wrong that NASA is receiving multi-billion-dollar funding when so many people are suffering hardship at home. Brevard County GP Doctor Henry Jenkins says people don't have to look far from the Cape Kennedy Space Center to see poverty and hunger. He said: "The irony is so apparent here. We're spending all this money to go to the Moon and here, right here in Brevard, I treat malnourished children with prominent ribs and pot bellies. I do see hunger."

Although a recent survey by the State Welfare Department found that 7,000 people in Brevard County are eligible for federal food aid, they don't receive it because the county is one of the few in Florida not to have paid for such a programme, meaning the poor have to rely on state welfare only. Jenkins complained: "They haven't addressed themselves to this problem. I guess they're more concerned about promoting tourism than caring about hungry people."

Single mother Lucille Wade, who lives in a low-rent housing district of West Cocoa, said: "It's rough, very rough. I've got six children, aged ten years to seventeen months, and there have been times when I've had to leave the house and look for work and they were waiting for me to come back with food. I get $152 a month from welfare. I was working at the Cape Kennedy Hilton as a maid, but there was a big lay-off and I never got my job back. The first of the month we do okay, but toward the end of the month I go crazy."

Dora Kimbrough, another mother of six from the same neighbourhood, said her husband earned $70 a week as a labourer, but explained: "We pay $25 a week rent and $23 for food. The rest goes on hospital bills. Friday is best because my husband gets paid and we eat good food. On Thursdays I pray for the best."

The County Welfare Department spent $103,118 last month on state welfare, helping 3,558 people. It spends another $55,000 a month on food aid. Edna Stoltz, County Welfare Programme Director, insisted: "Nobody's going to go hungry and the people in need are helped. Malnutrition, I would say, is at a minimum. To my mind, the main problem is doing some of the right things with the money they have. It's a matter of education and self-discipline."

Introducing a federal food distribution scheme would cost $80,000 a year but would bring in $828,900 worth of butter, cheese, flour, meat, milk and other staples for the poor. Of Florida's 67 counties, 51 have such a programme.

Richard Muldrew, County Commission Chairman, promised: "I know some people are hungry and we're going to do everything to end it here." But Lori Wilson, County Commissioner, admitted: "We can't just pull money out of thin air."

PAINTING HISTORY

Fourteen artists are busy producing a unique historical record of the Moon mission.

They have been commissioned to produce Apollo 11-related paintings as part of NASA's long-running 'Eyewitness to Space' scheme.

The project was the brainchild of NASA's first Administrator, James Webb, and has run since 1963 in conjunction with the National Gallery of Art in Washington. To date it has produced more than 500 sketches and paintings, which are currently being shown in an exhibition touring the USA.

Doctor Hereward Lester Cooke, National Gallery Curator of Painting, said: "The camera records every nut and bolt of the mission. We can't add to that. What we *can* add is emotional impact. What if an artist had been there when Christopher Columbus hoisted his sails?"

The artists are given free rein to observe and paint any part of the mission. Among the subjects chosen this time are the launch site, the splashdown operation and the astronauts themselves.

One artist taking part is Peter Hurd, who earned nationwide publicity two years ago when he painted a portrait of Lyndon Johnson. It was meant to be hung in the White House, but the President called it "the ugliest painting in the world" and refused to accept it. Hurd put it on display at a museum in Johnson's native Texas, attracting record attendances.

THE WRIGHT SPOT

People will be able to watch the Moon landing at the historic site where the Wright brothers flew the world's first powered aircraft.

A giant screen is being erected at Kill Devil Hills, North Carolina, in front of a monument marking Orville and Wilbur Wright's 1903 flight. The site will be open throughout the night next Sunday.

CHAPTER 14

MONDAY, 14 JULY
Launch Minus Two Days

"We certainly are thinking positively. I think we're quite well-suited to say 'when' we land, not 'if'."
> Apollo 11 Lunar Module Pilot Buzz Aldrin

"I've lost 12lb and I've stepped on a few snakes, but I'm the sort of bloke who doesn't like to admit defeat."
> Houston-To-Cape Kennedy Marathon Runner Bill Emmerton

Deke Slayton (seated far left) looks on as the astronauts (left to right) Buzz Aldrin, Neil Armstrong and Michael Collins give today's final pre-flight press conference via a closed circuit TV link at their quarters at the Kennedy Space Center.

'WE HAVE NO FEAR'

Neil Armstrong tonight pledged the Apollo 11 crew will go into their mission with "no fear".

The astronauts gave their final pre-launch press conference tonight, speaking from their quarters at Cape Kennedy.

Asked whether they felt fear going into the mission, Armstrong said: "After a decade of planning and hard work, we're willing and ready to attempt to achieve our national goal. I wouldn't say fear is an unknown emotion to us. Fear is characteristic of a knowledge that there may be something you haven't thought of and feel you might be unable to cope with.

"All the work that goes into the preparation for a flight does everything it can toward erasing those kinds of possibilities. As a crew, we have no fear of launching out on this expedition. We must realise that things can go wrong. I'm sure that American ingenuity and craftsmanship have given us the best equipment available."

Tonight's thirty-minute conference was conducted via closed-circuit TV, with fifteen miles separating the astronauts and newsmen. Asked about their workload, Armstrong said: "It's certainly been a very hard preparation time. However, our pace certainly hasn't been unreasonable. We're not unduly fatigued. We're ready to fly."

Earlier, the astronauts had put in more hours working in simulators, but Armstrong conceded: "Our simulators are amazing devices. However, I'm quite certain that we'll find a lot of things that are different than they were in the simulators."

He batted away a question about Richard Nixon cancelling their pre-flight dinner and Frank Borman's criticism of Doctor Charles Berry. He said dining with Nixon would have been "a great pleasure", but added: "I wouldn't attempt to have judgments in preventative medicine."

BUZZ LANDING PLEDGE

Buzz Aldrin has told the American public they can have total confidence the landing will succeed.

Aldrin was asked at tonight's press conference if he was worried by people using the term 'when', rather than 'if', about the landing.

He said: "We certainly are thinking positively. Everything we've been doing has been very positive. I think we're quite well-suited to say 'when' we land, not 'if'. I feel we're seeing the type of reaction that many of us were hoping for; a very enthusiastic public that has great confidence in what we hope to carry out for that American public."

The lunar module has had limited flight time, but Aldrin said: "I feel very confident of this vehicle. It's done a very good job in the two previous flights and I certainly feel it's ready. There's nothing I can think of that concerns me. We'd always like to have a little bit more fuel, but that's a natural reaction of a pilot; to want a little bit of extra protection stashed away in the hip pocket. Certainly, throughout the descent we'll be very closely monitoring the fuel margins."

When Aldrin was asked if he'd be disappointed if the Soviets' unmanned probe returned samples to Earth before Apollo 11, he conceded: "I'm sure we all would be."

MOON SPEECH MYSTERY

Neil Armstrong insists NASA chiefs haven't told him what to say when he steps on to the Moon, but claims he hasn't had time to decide what his words will be.

Armstrong told tonight's press conference: "Certainly nothing has been prepared for me and our attention has been focused on how to do the job and how to do it best and not so much with what might be the emotions of the moment.

"I don't plan to guess at this point what my emotions might dictate. As a matter of fact, I think perhaps the highlight for those of us in the LM (lunar module) will probably be a successful touchdown. I really look forward to that the most at this time."

Armstrong again hinted that a rest period might be scrapped, bringing forward the Moonwalk. "I'll be surprised if I'm able to get sound sleep on the lunar surface. Fortunately we have adequate rest periods on both the night before and the night after."

COLLINS TV BLACKOUT

Michael Collins admits he'll be jealous of Americans glued to their TV sets during the Moonwalk…because he won't be able to watch it.

He told tonight's press conference he'll be happy to be "minding the store" as he orbits the Moon alone for twenty-eight hours.

But he added: "I do have one complaint. I'd like to point out to those of you in the TV business that I have no set on board and I'm going to be one of the few Americans who isn't going to see the EVA (extra-vehicular activity). So I'd like you to save the tapes for me, please!"

VON BRAUN: NOW WE PRAY

Doctor Wernher Von Braun has promised the public that everything possible has been done to make the mission a success.

The Saturn V mastermind, speaking at a press conference at the Marshall Space Flight Center today, said he was totally satisfied with the Apollo 11 hardware and preparations for the mission.

He said: "I have complete confidence. We've done our job, all of us in this programme. All we can do now is pray. There's always some room for error or equipment failure. I hope we're all mature enough to absorb the shock of having a mission not completely successfully executed. I don't think it would happen, but it might."

Asked to weigh up the significance of the mission, he claimed: "It compares with the event of aquatic life crawling on land for the first time millions of years ago."

Preparations for launch remain ahead of schedule and tomorrow technicians will begin the crucial job of fuelling the Saturn V.

MOON DUST FIRE FEAR

A former space agency official has accused NASA of hushing up the danger of Moon dust causing a deadly fire on the lunar module.

Paul Haney, NASA Public Affairs Director until earlier this year, told a British TV channel today: "Moon dust carried into the Moon ship could explode or burst into flames. While the possibility is remote, it has been enough of a nagging worry to trigger a number of studies in the past three years; some even before the Apollo 1 fire.

"It's a danger that has never been mentioned publicly by NASA. The moment of truth will come when the astronauts return to the Moon ship after walking on the surface. Scientists think it possible that eons ago the dust may have contained oxygen. When exposed to the super-charged oxygen atmosphere inside the Moon ship, the dust may begin to give off heat. This is the moment the scientists will be holding their breath.

The astronauts have fire extinguishers, but nobody knows just how effective they would be."

A NASA spokesman insisted: "Committees have looked into it. That's how we came to the conclusion it's unlikely to happen. It's not considered worth making a big blow-up about. The chance does exist, however. That's one of the reasons why the lunar module has fire-fighting equipment."

JAN FEELS THE TENSION

Neil Armstrong's wife admitted to pre-launch nerves as she arrived in Cape Kennedy tonight.

Jan Armstrong, the only one of the wives who will attend, travelled from Houston by private jet. Asked how she was feeling, she conceded: "I'm a little bit tense."

In Houston, Pat Collins made light of the media siege outside her home. Asked whether she was nervous, she said: "Not yet". She then added with a smile: "I like splashdown best!"

Pat will watch the launch on TV, but in the latest issue of *Life* magazine she admitted: "It's pretty funny to think of myself, sitting with my eyes glued to the set watching Mike start off toward the Moon. I'm the girl who used to sit and listen to the Mercury shots on the radio because I wasn't going to watch one of those things go to pieces right in front of me."

Collins' mother, Virginia, will watch the launch at home in Washington DC. She said: "I'm looking forward to it. I took special interest in Apollo 10 because I knew my son would be on the next one. I may be a little fearful – I think we're all a little fearful of the unknown – but I'll live!"

CAPE'S RING OF STEEL

A team of 500 security guards are making round-the-clock patrols at Cape Kennedy Space Center in the build-up to launch.

The sprawling site includes acres of swamps and dunes, plus twenty miles of beaches, so guards are patrolling on foot and in boats, helicopters and beach buggies. Some are equipped with the same $1,000 night-vision devices used by troops in Vietnam. Checks are being carried out on vehicles driving in and out of the site, which can be as many as 40,000 a day.

Charles Buckley, Security Director, said no unauthorised person had managed to gain access to the site for four years. He said: "The last time, during the Gemini 4 launch, six different people tried to swim into or walk into the launch area and were all caught.

"My men have to be on the alert, sometimes as much for the safety of intruders as protecting the launch facilities. The swampy area is filled with snakes and alligators and anyone who tried to get through it probably would have more trouble from them than from us."

Gordon Harris, NASA Director of Public Affairs, admitted, however, that it might prove impossible to stop a would-be saboteur. He said: "It would be very difficult, say, to prevent a frogman from landing on the beach near the launchpad and taking a shot at the vehicle with a high-powered rifle."

BATTLE OF THE 'BIRDWATCHERS'

Space enthusiasts were jealously guarding their territory today as they nabbed the best vantage points for launch.

Hundreds of thousands of visitors – dubbed 'birdwatchers' by locals – are already camped out in tents, caravans and campervans. Herbert Johnson, Director of Brevard County Civil Defense Agency, said: "People are pouring in. It seems to be the largest crowd we've ever had and, off-hand, I'd say we already have as many today as we had for Apollo 10."

At Port Canaveral Jetty Park, where visitors are paying $2-a-night, the Fields family from Columbus, Ohio, arrived early to secure their pitch. Pryse Fields, a crane operator, said: "We've seen four space shots from this spot, but this will be the first manned one we've seen. I'm afraid to leave this spot; someone else would get it."

Titusville beach offers a direct view of the launch tower and among those who have their telescopes trained on the Saturn V is Dennis Donnelly, 27, from Albuquerque, New Mexico. He has a home-made wood-framed telescope and explained: "I finished it three weeks ago. It's real nice on a clear day. I can see people walking on the gantry of the launch tower."

Thousands have been taking the official space centre tour to get a closer look at the inner workings of the Apollo operation. Roy Childress, from Elizabethtown, Tennessee, said: "The whole darned thing is just beyond my comprehension. It's just fabulous."

With around one million people expected to be in Brevard County to watch the launch, many are camping out on beaches near the Kennedy Space Center.

Five planeloads of European tourists arrived today on a TWA package trip. Retired French postal inspector Marcel Bendreham said: "I want to see the launch because it's such a sensational event."

HE'S THE PARTY KING

Dozens of astronauts and their wives were among the guests at a lavish boat party thrown tonight in Cocoa Beach by oil and gas tycoon John McCandish King.

Invites to Colorado businessman King's pre-launch parties have become the hottest ticket in town since he started them during the Gemini programme. He said: "I guess it's on in a big way; there are parties everywhere. A few years ago, when we started the parties, it used to be a little-bitty thing. I don't know what I'll do if it gets any bigger!

"I call these people 'The Loyalties'. They're the ones who've been in on this thing for a long time. This is kind of an historic time for these people. Lots wish they were going to be the guys going up on that firecracker."

Guests at a reception held today by CBS included science fiction writer Arthur C. Clarke, whose short story *The Sentinel* inspired Stanley Kubrick's hit movie *2001: A Space Odyssey*. The film has grossed millions since its release last year and Clarke admitted he has watched it sixteen times. He said: "You have to see it three times just to get the impact. The ending is better than smoking pot … although I've never tried that!"

PUBLIC BACK MISSION

A new poll suggests public support in America for the mission has increased now that the launch is imminent.

The survey by Louis Harris & Associates shows that fifty-one per cent are in favour, with forty-one per cent against. In a similar Harris poll in February, forty-nine per cent said they didn't agree with the mission and only thirty-nine per cent backed it. Louis Harris said the result reflected the fact people were thinking "if we've gone this far we ought to finish the job".

NATIONS' MESSAGES REVEALED

Seventy-one nations have sent messages for inclusion on a disc to be left on the Moon.

NASA today disclosed details of goodwill statements submitted from dozens of political figures, religious leaders, including the Pope, and heads of state such as Britain's Queen Elizabeth II. The USSR and China have chosen not to be represented.

The quirkiest message is from Felix Houphouet-Bolgny, Ivory Coast President, who asks the astronauts "to tell the Moon how beautiful it is when it illuminates our nights".

AMERICA 'WILL BURN'

American cities will "burn down" if politicians don't tackle poverty and hunger.

That was the stark warning today from the Southern Christian Leadership Conference (SCLC) as it transported twenty-five African-American families from five southern states to Cape Kennedy for a protest march.

Reverend Hosea Williams, Director of the group's Poor People's Campaign, said: "This isn't about our ability to explore outer space. We're protesting about America's inability to choose proper priorities."

Walter Fauntroy, Director of the SCLC's Washington bureau, said the astronauts would be eating meals costing up to $300 each during what he described as their "joyride", while poor people were living on eighty cents a day.

He said: "The space programme doesn't deal with the problems of American society. If they don't do something about poverty and hunger, the country is going to burn down."

MAKING MAPS TOOK FIVE YEARS

The astronauts will have nearly a hundred lunar maps to help them get their bearings on the Moon.

It has taken five years of painstaking work to piece together the maps using photos and TV footage taken by twenty-one different satellites, some of which landed on the Moon.

Revealing details of the map package, Doctor Thor Karlstrom, US Geological Survey Mission Planning Co-ordinator, explained that, for landing, there were five large-scale maps highlighting key landmarks and hazards such as large craters and boulders.

To orientate themselves on the surface, the astronauts will have ninety-two detailed maps of 'Landing Site 2', a relatively flat area of about thirty-six square miles in the south-eastern corner of the Sea of Tranquility.

While planning the mission, NASA has been able to study three-dimensional views of the landscape put together by scientists using techniques similar to those employed in Hollywood to make 3D movies.

ALASKA IN THE PICTURE

People living in Alaska have won their battle to have live TV coverage of Apollo 11.

The state, in the remote north-west of America, isn't connected to national networks because it has no satellite service and insufficient telephone cables. Shows are taped and mailed to Alaska's biggest city, Anchorage, for delayed screening.

However, Senator Mike Gravel has persuaded the Department of Defense to rig up a network using civilian and military satellites to relay pictures to Earth, bounce them into space again and then back to Anchorage.

A spokesman for the Senator said: "It's the greatest thing since statehood. People are swarming into Anchorage for live TV. Every hotel room is booked. You have to remember, Alaska is the place where people normally see Christmas shows in March!"

MARATHON MAN

Marathon man Bill Emmerton arrived in Cocoa Beach today after running 1,034 miles from the Manned Spacecraft Center in Houston.

The Australian, a professional runner who now lives in America, made the run in a Rotary Club-backed initiative to encourage people to take up jogging to keep fit. He admitted today: "I'd rather go to the Moon every week than do it again!"

Emmerton, 48, was given a NASA medical before setting off on 17 June and was greeted today by Apollo 8 hero Jim Lovell, a Presidential Adviser on physical fitness.

His previous feats have included running through Death Valley in California and Nevada. Emmerton said: "I'd rather go through Death Valley again than repeat this. I've lost 12lb and stepped on a few snakes. Yes, I'm suffering and I'm in pain, but no one else can withstand it as I do. I'm the sort of bloke who doesn't like to admit defeat."

Jogging has yet to catch on in the USA and Emmerton hopes his efforts will encourage people to take it up. He lives in Los Angeles, California, where there is a large hippy community. He added: "I don't know what's going to happen to those sorts of people. They're like sheep lost in a storm, wandering around, not knowing where to go, what to do. Maybe I can set a better example."

CHAPTER 15

TUESDAY, 15 JULY
Launch Minus One Day

"I've never seen calmer men. You'd think they were getting ready for a day's golf instead of the Moon ride."
    Kennedy Space Center Chief of Public Information Jack King

"Men will land from Russia and the United States, each party complete with H-bomb and each intent upon exterminating the other. It would be cheaper to shoot each other at home."
    Philosopher Lord Bertrand Russell

Buzz Aldrin (left) and Neil Armstrong, pictured rehearsing working on the lunar surface in a simulated Moon environment, will be doomed to die if the lunar module crash lands or fails to take off.

SUICIDE PILLS VETOED

Space agency chiefs today ruled out issuing suicide pills for Neil Armstrong and Buzz Aldrin to take if they are stranded on the Moon.

If the lunar module crash-lands or its engine fails on take-off, the pair will be doomed to die when their oxygen runs out. Doctor Chris Kraft, Director of Flight Operations, said: "I don't even like to think about it, because there really isn't anything else that could be done."

Asked by a reporter what would be done to "ease the astronauts' suffering", Doctor George Mueller, Associate Administrator for Manned Space Flight, said: "There's no rescue capability. We've based everything on making their vehicle work. There are no suicide capsules or anything of that sort. Nor would they plan to take advantage of such things."

NASA says the astronauts could survive for as long as a week, but that there is no way they could be reached in time; even if Apollo 12 were launched on a rescue mission.

Countdown remained ahead of schedule today, allowing an eleven-hour 'hold' period for staff to take a break. With steel scaffolding removed from the rocket stack, the Saturn V and spacecraft could be seen on the floodlit launchpad from up to thirty miles away. Technicians were tonight chilling the Saturn V's tanks ready to pump in super-cold fuel at 10,000 gallons a minute.

In a reference to the growing jams on local roads, Doctor Kurt Debus, Kennedy Space Center Director, joked: "There's nothing I know of that would prevent us from launching on time. The only thing we haven't simulated is the traffic!"

Paul Donnelly, Launch Operations Manager, reported: "So far, it's been the smoothest Apollo countdown we've had."

Explaining how the final minutes before lift-off will unfold, Donnelly said it was "an old wives' tale" that someone in the firing room pressed a button to start the engines. Instead, a computer automatically starts the sequence with three minutes, ten seconds to go. With 8.9 seconds left, ignition of the first-stage will begin and, two seconds before lift-off, all five engines should be running. The arms holding the rocket will only be released at the last moment, ensuring the engines reach their full thrust.

Neil Armstrong (left), Michael Collins (centre) and Buzz Aldrin pose for an official pre-mission portrait.

The Lunar Landing Research Vehicle, nicknamed the 'Flying Bedstead', which Neil Armstrong flew to help prepare for the Moon landing. Armstrong cheated death in 1968 when he parachuted to safety moments before the contraption crashed and exploded.

The fully-assembled Apollo 11 stack inches its way out of the giant Vehicle Assembly Building on a huge crawler transporter to make the journey to the launchpad.

Left: The giant Saturn V first stage engines, which made the Apollo missions possible, on display at the Kennedy Space Center in Florida.

Below: A NASA illustration from 1971 explaining the specifications of the Saturn V launch vehicle – the three stages of the rocket, the instrument unit and the Apollo spacecraft.

SATURN V LAUNCH VEHICLE

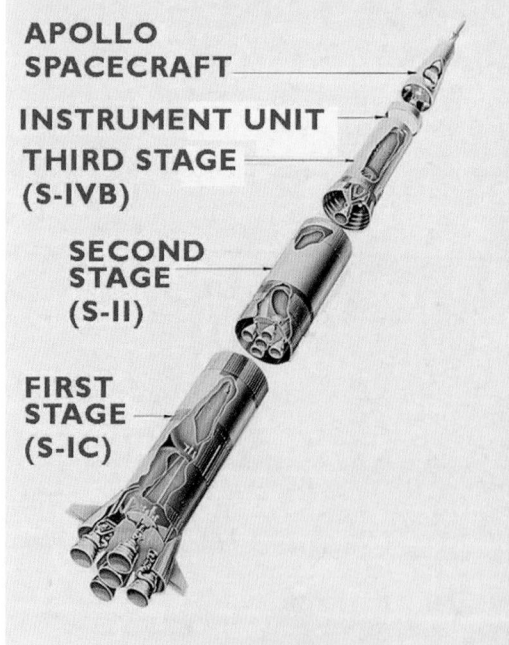

APOLLO
SPACECRAFT

INSTRUMENT UNIT

THIRD STAGE
(S-IVB)

SECOND
STAGE
(S-II)

FIRST
STAGE
(S-IC)

MSFC-71-PM 1200-127

CHARACTERISTICS

LENGTH (VEHICLE)...86m
　　　(VEHICLE, SPACECRAFT, LES)...111m
WEIGHT　(TOTAL DRY).......................178,822kg
　　　　(TOTAL WET)................2,708,831kg
　　　　(AT LIFTOFF)...............2,903,000kg
TRANSLUNAR PAYLOAD CAP..............48,500kg
EARTH ORBIT PAYLOAD (2 STAGE).....96,000kg

STAGES

FIRST (S-IC)
SIZE...10 x 42m
ENGINES...5 F-1
THRUST.....................................3,470,000kg
PROPELLANTS WEIGHT (LOX)......1,497,856kg
　　　　　　　　　(RP-1)...........651,500kg

SECOND (S-II)
SIZE..10 x 24.8m
ENGINES..5 J-2
THRUST..526,176kg
PROPELLANTS WEIGHT (LOX)..........379,339kg
　　　　　　　　　(LH2)............72,387kg

THIRD (S-IVB)
SIZE...6.6 x 18.1m
ENGINES...1 J-2
THRUST..104,328kg
PROPELLANTS WEIGHT (LOX)............86,000kg
　　　　　　　　　(LH2)............19,700kg

INSTRUMENT UNIT
SIZE...6.6 x .91m
WEIGHT..2,038kg
GUIDANCE SYSTEM..........................INERTIAL

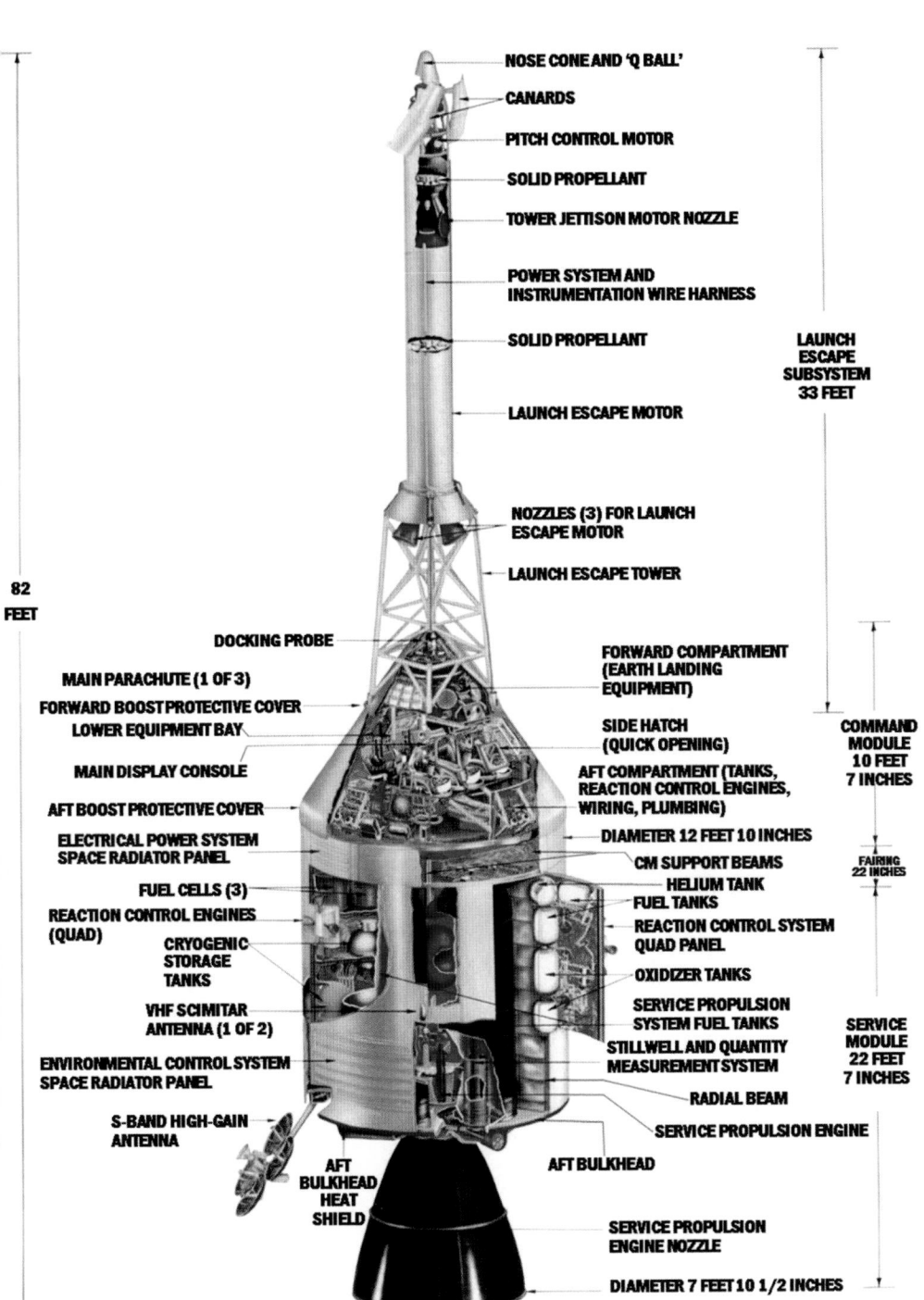

NOSE CONE AND 'Q BALL'

CANARDS

PITCH CONTROL MOTOR

SOLID PROPELLANT

TOWER JETTISON MOTOR NOZZLE

POWER SYSTEM AND
INSTRUMENTATION WIRE HARNESS

SOLID PROPELLANT

LAUNCH ESCAPE MOTOR

LAUNCH ESCAPE SUBSYSTEM 33 FEET

82 FEET

NOZZLES (3) FOR LAUNCH ESCAPE MOTOR

LAUNCH ESCAPE TOWER

DOCKING PROBE

MAIN PARACHUTE (1 OF 3)

FORWARD BOOST PROTECTIVE COVER
LOWER EQUIPMENT BAY

MAIN DISPLAY CONSOLE

AFT BOOST PROTECTIVE COVER

ELECTRICAL POWER SYSTEM
SPACE RADIATOR PANEL

FUEL CELLS (3)

REACTION CONTROL ENGINES
(QUAD)

CRYOGENIC
STORAGE
TANKS

VHF SCIMITAR
ANTENNA (1 OF 2)

ENVIRONMENTAL CONTROL SYSTEM
SPACE RADIATOR PANEL

S-BAND HIGH-GAIN
ANTENNA

AFT
BULKHEAD
HEAT
SHIELD

FORWARD COMPARTMENT
(EARTH LANDING
EQUIPMENT)

SIDE HATCH
(QUICK OPENING)

AFT COMPARTMENT (TANKS,
REACTION CONTROL ENGINES,
WIRING, PLUMBING)

DIAMETER 12 FEET 10 INCHES

CM SUPPORT BEAMS

HELIUM TANK
FUEL TANKS

REACTION CONTROL SYSTEM
QUAD PANEL

OXIDIZER TANKS

SERVICE PROPULSION
SYSTEM FUEL TANKS

STILLWELL AND QUANTITY
MEASUREMENT SYSTEM

RADIAL BEAM

SERVICE PROPULSION ENGINE

AFT BULKHEAD

SERVICE PROPULSION
ENGINE NOZZLE

DIAMETER 7 FEET 10 1/2 INCHES

COMMAND
MODULE
10 FEET
7 INCHES

FAIRING
22 INCHES

SERVICE
MODULE
22 FEET
7 INCHES

An illustration produced by contractor North American Rockwell in 1968 explaining
the details of the command and service modules and the launch escape system.

Deke Slayton, Director of Flight Crew Operations (far right), reviews lunar charts with the crew as they have breakfast on the morning of launch. Buzz Aldrin is seated next to Slayton and Michael Collins is opposite him. Next to Collins is Neil Armstrong, with Apollo 8 Lunar Module Pilot Bill Anders also joining them for a traditional pre-flight meal of steak and eggs.

Neil Armstrong waves to staff in a corridor of the Kennedy Space Center as the astronauts leave to board the spacecraft on launch day.

Above: Rows of flight technicians at Launch Control at the Kennedy Space Center in Florida.

Below left: Apollo 11 lifts off at 9.32am EDT on 16 July 1969.

Below right: A close-up view of the commemorative plaque attached to the lunar module descent stage.

Apollo 11 blazes into the distance with the US flag in the foreground.

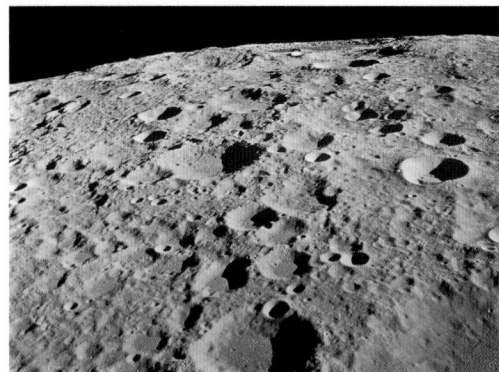

A detailed view of the far side of the Moon taken from Apollo 11 while in lunar orbit.

Buzz Aldrin, photographed by Neil Armstrong, inside the lunar module on the day of the landing.

Buzz Aldrin on the lunar surface. Neil Armstrong, who took the photograph, can be seen reflected in Aldrin's helmet visor.

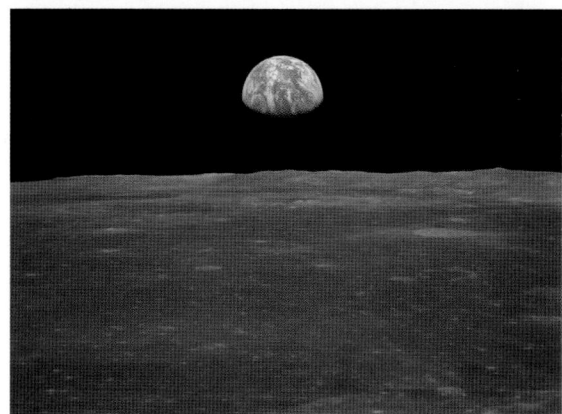

A photograph taken from the Apollo 11 command module shows the Earth rising beyond the lunar horizon.

The lunar module, photographed from the command module, begins its descent to the lunar surface after separation.

The astronauts relax in the Mobile Quarantine Facility on their journey back to the USA. Neil Armstrong is in the foreground with Michael Collins on the left and Buzz Aldrin on the right.

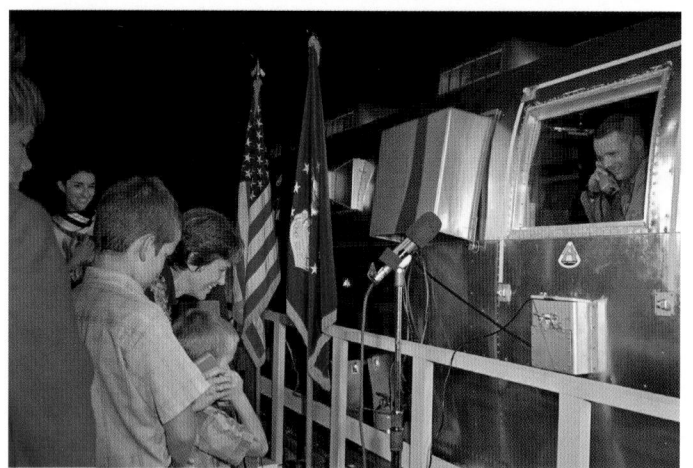

Neil Armstrong speaks to his youngest son, Mark, via an intercom while wife Jan and oldest son Ricky look on. The families were waiting to greet the crew when the Mobile Quarantine Facility arrived at Ellington Air Force Base, Texas, on 27 July 1969.

The Moon men are given a heroes' welcome in New York, with tickertape showering down as they are driven along Broadway and Park Avenue.

CREW 'SLEPT LIKE BABIES'

The astronauts were showing no sign of nerves today as the clock ticked down to their historic flight.

Neil Armstrong, Buzz Aldrin and Michael Collins had one last session in flight simulators, but spent most of the day relaxing in their quarters and reviewing the flight plans.

Jack King, Kennedy Space Center Chief of Public Information, said: "I've never seen calmer men. They slept like babies last night. You'd think they were getting ready for a day's golf instead of the Moon ride."

The astronauts phoned their loved ones and Armstrong's father, Stephen, revealed: "Neil's more calm and collected than his Mom and Dad."

Rather than dining with Richard Nixon tonight, as originally planned, the crew ate steak and mash with seven other astronauts before going to bed at around 9pm.

The crew enjoyed a pre-flight dinner of steak and mashed potatoes this evening with seven other astronauts.

The President did phone the crew to wish them luck. Ronald Ziegler, White House Press Secretary, said Nixon told them: "What you're doing will lift the spirits of the American people, as well as the world. You carry with you a feeling of goodwill in this greatest adventure man has ever seen."

BASEBALL 'IN ORBIT'

Baseball star Tim Cullen thrilled Richard Nixon and astronaut Frank Borman with his own 'space launch' tonight.

Nixon and Borman were at Robert F. Kennedy Stadium to see the President's favourite team, the Washington Senators, beat the Detroit Tigers. Senators infielder Cullen hit his first home run of the year and said: "I knew President Nixon was here. Someone told me Frank Borman was in the stands, too, and as I rounded second base after hitting the homer I was thinking 'I guess I showed him how to put the ball in orbit'."

SPACE BOSS MEETS PROTESTORS

America's top space boss has told hunger marchers he'd happily scrap the Moon shot if he thought it would stop poverty and hunger.

Reverend Ralph Abernathy, President of the Southern Christian Leadership Conference (SCLC), led a 150-strong protest at Cape Kennedy today.

Doctor Thomas Paine, NASA Administrator, met them outside the space centre gates and said: "It'll be a lot harder to solve the problems of hunger and poverty than it is to send men to the Moon. If it were possible for us not to push that button tomorrow and solve the problems you're talking about, we wouldn't push the button."

Paine – who issued forty VIP launch passes to the protestors – added: "We feel the space programme is a programme for all Americans. We don't regard it as a programme against the anti-poverty programme. I want you to hitch your wagon to our rocket and tell the people that the NASA programme is an example of what this country can do."

Earlier, during a rally at a church in West Cocoa, Abernathy insisted: "I'm proud of those astronauts. I'm happy we're going to the Moon. I'd be just a little more happy if we learned how to live down here on Earth. America has mixed-up priorities."

ONE MILLION SET TO WATCH

Half a million tourists had arrived in Brevard County by sunset today; 200,000 more than the record number who witnessed the launch of Apollo 10.

A thirty-five-mile stretch of Highway US-1 leading into the area was jammed with traffic and Herbert Johnson, County Director of Civil Defense, predicted: "Based on motel bookings, our traffic count, the airports and beaches, we expect to reach the million by eight o'clock tomorrow morning."

As around 1,000 highway patrolmen, sheriff's deputies and state troopers tried to keep the traffic flowing, Leigh Wilson, County Sheriff, said: "Right now, we have more people than we've ever had. I'm not sure our roads can handle it."

Jimmy Blount, a 21-year-old University of Georgia student, is camped at Titusville, eleven miles to the west of the floodlit Saturn V. He said: "It looks like one big beautiful white candle. I'm going to tell my grandchildren about it; 'I was there when men went to the Moon'."

Engineering planner William Nelson, from Durham, Connecticut, is camped out with his family. He said: "They tell me I'll be able to feel the Earth shake when it goes off. Once I see it, I'll know that it was worth all the heat and mosquitoes here."

Cal Rogers, an Oklahoma senior citizen, said: "This is it! This is what we've been working and paying for for so long."

WORLD PEACE HOPE

Lyndon Johnson today made a rousing plea for America to use the Moon shot as an inspiration to achieve peace.

Johnson – whose five-year term as President was blighted by civil unrest over racial and social issues and the Vietnam War – was one of the first VIP guests to arrive in Cape Kennedy, flying in from Texas on a military jet.

Speaking at a Cape Kennedy Hilton lunch held by the Gannett Newspaper Group in honour of Johnson and James Webb, former NASA Administrator, the ex-President said: "If we can lead the world to the Moon, we can lead the world to peace and bountiful prosperity here at home."

Referring to civil unrest, Johnson quoted a remark made by Benjamin Franklin after the signing of the Declaration of Independence. Franklin, asked whether the United States was a republic or a monarchy, replied: "We've given you a republic … if you can keep it." Johnson said today: "In this day of rebellion against authority, we do have a problem of keeping it. I think if we can go to the Moon, we *can* keep it."

STORM BATTERS HOUSTON

Michael Collins' family had a pre-launch scare today when a large tree was blown down outside their home in storm-hit Texas.

Winds gusting at more than 40mph battered the Houston area, including NASA's Manned Spacecraft Center and the Nassau Bay suburb where the Collins live.

A 30-ft hackberry tree cracked about 2ft above the ground and was sent tumbling on to the Collins' front lawn. After emerging to inspect the damage, Pat Collins said: "It was quite a shock. We heard it cracking. It snapped like a toothpick."

NASA, which arranged for the fallen tree to be removed, also reported the storm had caused minor damage at the space centre, but nothing that would impact on the mission.

SWIMMING POOL ON HOLD

Space agency chiefs swooped into emergency action today…to stop Buzz Aldrin's neighbour building a swimming pool.

A NASA liaison official at the Aldrins' home noticed workmen had arrived next door to start digging out the backyard. He persuaded the neighbours to postpone the job to avoid the chaos of lorries trying to access the cul-de-sac when a mass of media trucks are already parked there.

SPACE WAR WARNING

Philosopher Lord Bertrand Russell today claimed man should stay away from the Moon or risk a world war in space.

The 97-year-old Welsh aristocrat, a long-time peace campaigner who served a prison term for promoting pacifism during World War I, believes the space race has been driven by "ruthless competition" rather than a spirit of adventure.

Writing in British newspaper *The Times*, the former Nobel Literature Prize winner warned: "Men won't be content to land upon the Moon and try to make it habitable. They will land simultaneously from Russia and the United States, each party complete with H-bomb and each intent upon exterminating the other. It would be cheaper to shoot each other at home.

"I should wish to see a little more wisdom in the conduct of affairs on Earth before we extend our deadly disputes to other parts. It is for us to grow to the stature of the cosmos, not to degrade the cosmos to the level of our futile squabbles."

Russell recalled how he was captivated as a boy by Jules Verne's fictional Moon exploration story, but added: "The thought of such an adventure is exciting, especially to those who are still young, but those who are no longer young are troubled by doubts.

"Unfortunately, the spirit of ruthless competition has infected projects for reaching the Moon. It was felt that the important thing wasn't that the Moon should be reached, but that it should be reached by 'our side' – whichever that may be – sooner than by the other. This makes the whole enterprise one in which it is difficult for sane men to see much of value."

CATCH THE SUN

A scientist explained today how the astronauts will try to catch some sun while walking on the Moon.

Buzz Aldrin will set up an experiment devised by Swiss scientists, with a sheet of aluminium foil being used in an attempt to capture tiny particles carried ninety-one million miles to the Moon by solar winds.

Doctor Johannes Geiss, of the University of Berne, said Aldrin will unfurl the 1ft-by-4ft sheet on the surface like a piece of kitchen foil. It will be retrieved before the astronauts leave the surface.

Geiss said he hopes to find traces of noble gases like neon, argon, krypton and xenon. "We hope a tiny amount of solar material, perhaps one-billionth of an ounce, will be in the foil. There may be surprises that no one has anticipated. The experiment may give us a way of observing the process of meteoritic formation and, through that, perhaps of the formation of the solar system."

SOVIET FLIGHT IS 'UNIQUE'

A space expert today revealed that Luna 15 is on a different flight path to previous Soviet Moon probes, suggesting it *will* attempt a landing.

Sir Bernard Lovell, Director of the Jodrell Bank Observatory in Cheshire, England, said the craft was travelling slower, possibly to save fuel to brake for a soft landing and a take-off from the surface.

He explained: "This is unique as far as Luna missions are concerned. We simply don't know what Luna 15 is going to do, but there is every indication that it may be attempting something new. It supports the theory that Luna may be attempting to recover lunar rock. It would be a remarkable achievement."

Lovell believes the race to the Moon has the potential to begin a phase in history with "immense possibilities for good or evil". He said: "The time will come, within ten years, when lunar bases are established. International co-operation will become essential, otherwise a very serious situation might arise, both scientifically and politically."

However, Lovell dismissed criticism of the USA's space spending, adding: "The annual cost is only a third of the annual tobacco bill of the American people and one-sixth of their drinks bill. Arguments about the national economy are convenient excuses for a lack of vision."

'SERIOUS QUESTIONS' AHEAD

Future spending on space will come under "serious question" from the American people after the Moon landing, a NASA official admitted today.

What direction the programme should take is currently being considered by a panel tasked with making recommendations to Richard Nixon.

And John Hodge, Advanced Missions Programme Manager, said: "It's a very serious question that we in NASA have to face. We're trying to put together a plan. Of course, it's then up to the nation – and Congress, in particular – to decide how much it wishes to support it.

"People talk about the returns of the space programme as being Teflon frying pans and that kind of stuff, but the real return is a basic capability that is developed within the country. It's also a peaceful way of waging the technology game. In the past, technology has always been driven by war. These are real advantages. If, on top of that, you develop a degree of national prestige, you're getting a two-fold return."

He added: "One of the things we'll be doing in these early lunar flights is to find a good place for a Moon base. Given the money, it would be quite feasible to have it by the end of the 1970s. If you look farther, I'm sure we'll want to go to the planets. It's not unreasonable to think that we'll have landed on Mars before the end of the century."

Doctor Thomas Paine, NASA Administrator, knows he faces a tough job convincing Congress to back future spending, but he has no regrets about leaving a successful career with General Electric eighteen months ago to head the space programme. He said: "I liked my years with General Electric very much, but the question is whether you want to devote your life to devising the perfect watt-hour meter or to the electric toothbrush market. I decided the time had come to accept a new challenge."

ELEPHANT IN SPACE

A book telling the Apollo 11 astronauts' exclusive story is set for some serious competition…from *Babar The Elephant*.

A surge in space-related books will continue with the release of dozens more in the next few months. A *Time-Life* book exclusively recounting the Apollo 11 story in the astronauts' own words is expected to be the biggest seller.

But industry experts believe a new pop-up book in the popular children's series, *Babar,* will be one of the biggest hits. *Babar's Moon Trip* will see King Babar and his wife, Queen Celeste – wearing elephant-sized space helmets – become the first animals to land on the Moon. *Babar* has been popular worldwide since Frenchman Jean de Brunhoff's first book in 1931. His son, Laurent, has continued the series since his father's death.

The first Apollo 11-related title to hit the shelves, though, will be *New York Times* aerospace writer John Noble Wilford's account of the mission. He will write the final chapter of *We Reach The Moon* within twenty-four hours of splashdown, ready for an immediate print-run of 300,000 copies.

CHAPTER 16

WEDNESDAY, 16 JULY
Launch Day

"They're ready to go to work, the same as any other morning."
Director of Flight Crew Operations Deke Slayton

"Oh man! That was beautiful...the greatest thing I've ever seen."
Cape Kennedy Launch Tourist Gerald Siegelin

Apollo 11 blasts off today to launch Neil Armstrong, Buzz Aldrin and Michael Collins on their historic mission.

'A MAGNIFICENT RIDE'

Apollo 11 was hurtling towards the Moon tonight after a near-flawless first day of its historic mission.

After a spectacular launch from Cape Kennedy, every stage went like clockwork as the spacecraft achieved Earth orbit and then set a course for the Moon.

A worldwide TV audience of hundreds of millions watched spellbound as the countdown clock reached zero and the Saturn V's engines reached full thrust, engulfing the base of the 363ft-high stack in flames.

Up to a million people were packed into the area around Cape Kennedy and they felt the ground shake beneath their feet as the 3,200-ton rocket strained to break free of Launchpad 39A. As the launch tower arms fell away, the monstrous rocket slowly and majestically heaved itself off the ground.

Lift-off came as scheduled at 9.32am and when – two hours and fifty-three minutes later – the third stage had fired Apollo out of Earth orbit, Mission Control told the crew: "It looks like you're well on your way now." Neil Armstrong reported: "That Saturn gave us a magnificent ride. It was beautiful."

Medical data showed that, during the first twelve minutes, Armstrong's heart rate got no higher than 110bpm. Michael Collins' peak was 99bpm and Aldrin's just 88bpm.

The trio had been woken up at 4.15am and given a brief final medical examination before enjoying a steak and scrambled eggs breakfast. Shortly before 7am, hundreds of staff assembled to wave them off on the eight-mile drive to the launchpad and Deke Slayton, Director of Flight Crew Operations, said: "They're up and charging. They're ready to go to work, the same as any other morning. There was normal flight chatter at breakfast. They're calm, cool and collected."

Once they were settled into the command module, Guenther Wendt, Launchpad Close-out Crew Leader, gave them a cheery wave and sealed the hatch.

In temporary grandstands outside, VIP guests, headed by Vice President Spiro Agnew and former President Lyndon Johnson, were kept up to date by a running commentary over the public address system from Jack King, Chief of Public Information at Cape Kennedy. As temperatures soared, King told them: "It's a beautiful morning for a trip to the Moon."

His constant updates built the excitement. "T-minus thirty minutes, fifty-two seconds and counting. Countdown proceeding very satisfactorily [...] T-minus twenty-one minutes, fifty-five seconds and counting. All weather conditions are acceptable for launch [...] T-minus ten. We're aiming for our planned lift-off at thirty-two minutes past the hour."

The astronauts endured a two-hour wait inside the capsule, but there was a break in the tension when Paul Donnelly, Launch Operations Manager, told them preparations were fifteen minutes ahead of schedule and Armstrong joked: "That's fine. Just as long as we don't launch fifteen minutes early!"

As launch neared, King announced: "T-minus one minute, thirty-five seconds on the mission to land the first men on the Moon. All indications at this time indicate we are go!" Finally, picking up the countdown, King told the crowd: "Twelve, eleven, ten, nine ... ignition sequence start ... six, five, four, three, two, one, zero ... all engines running." Flames and smoke spewed from the first-stage engines and the roar sent thousands of birds scattering in fright.

The crowd stood almost as one as the Saturn V began to heave itself off the ground and King confirmed: "We have lift-off at thirty-two minutes past the hour ... lift-off on Apollo 11!"

There were cheers and applause, although many people stared in stunned silence for the three minutes they could see Apollo 11 riding into the distance on a bright orange flame. TV commentators struggled for the right words. Eric Sevareid of CBS said: "We don't have language to adequately describe this event. How can you say 'As high as the sky' or 'Reach for the sky' now?"

When it was over, VIP guests spoke of their excitement. Secretary of Commerce Maurice Stans said: "What a great experience! Too thrilling for words." British science fiction writer Arthur C. Clarke admitted: "I haven't cried or prayed for twenty years, but I did both this morning. This is the last day of the old world." TV star Ed McMahon – Johnny Carson's sidekick on NBC's *The Tonight Show* – said: "I've never seen anything like it. As an old Marines pilot, I can't believe what I saw."

Around 250 Congressmen and Senators attended and George Miller, Chairman of the House Committee on Science and Astronautics, said: "This is the greatest day for all of us; for all the world."

Former President Johnson said one thing he thought about during launch was "something you don't hear too much about these days, a great pride in this country". He said: "It seemed like the half a million people who worked on this programme were here lifting this thing."

NATION AT STANDSTILL

America came to a virtual standstill as people followed the launch on TV and radio.

In a scene repeated all over the country, dozens in Fredericksburg, Virginia, watched on a colour TV displayed in a shop window. One local, Chuck Bernard, said: "Oh, baby! That's the most! When they land on the Moon I'll scream!"

At casinos in Las Vegas, Nevada, gamblers deserted the twenty-four-hour gaming tables to watch. In San Francisco, California, where it was 6.32am, the open-air fish market was unusually silent as workers listened to radio commentary.

Angus MacDonald, 20, was among Air Force Academy cadets watching in Colorado Springs at 7.32am and said: "As the spaceship lifted off, we began to cheer and clap and yell and scream. It was like our team winning a football game. All I could think about was myself and Mars; I want to be the one to go there. I *have* to be."

Cowhands at a remote Wyoming ranch, where there's no radio or TV signal, interrupted their work at 7.32am to honour Apollo 11. Ranch owner Doctor Oakleigh Thorn explained: "We feel so close to the Moon shot out here because we're so close to the stars and sky. Last night we sat under the trees and looked at the Moon. We talked about how extraordinary it is that man is going to that white, white place which has always seemed so untouchable and remote."

HEAVEN AND HELL

Record crowds marvelled at Apollo 11 soaring into the heavens – then set off on a hell ride home.

With Brevard County roads at a virtual standstill, many motorists watched the launch from the roadside, with some standing on car roofs. Traffic was so bad overnight that many motorists had pulled over and slept in their vehicles, ditches or fields.

After launch, traffic on Highway US-1 was backed up for more than ten miles in what Sheriff's Deputy Coy Granger described as "a real serious tie-up".

Hundreds of thousands watched the launch from local beaches. Electrical engineer Gerald Siegelin, from Kingsville, Texas, slept on a grassy bank and woke up covered in ants. But the launch didn't disappoint. "Oh man! That was beautiful," he said. "Boy! That's the greatest thing I've ever seen!"

Paul Pearson, an electronics technician from Illinois, said: "I'm stunned. Once they land on the Moon, it will represent a quantum jump of mankind similar to the transition of animals from water to land, way before mankind developed. It's evolutionary."

Student Charles Walker, 20, marvelled: "It's like mankind has developed fire all over again. Perhaps this will be the kindling to bring men together."

Philip Buchanan, a 25-year-old student at Syracuse University, New York, insisted: "I've read that we ought to spend money on poverty, but I think this is just as important."

Teacher Joseph Bunecky, from Monroeville, Pennsylvania, said: "I wonder if Christopher Columbus had this kind of send-off?"

Many spectators gave up on trying to beat the traffic on the roads around Cape Kennedy this morning and pulled over to watch the launch from where they parked.

In Jules Verne's *From The Earth To The Moon*, the lunar flight was sponsored by the fictional Baltimore Gun Club. Today, three men from the Maryland city paid their own tribute as they watched from Port Canaveral. Carl Wasko, Robert Vicek and Robert Green have set up their own Baltimore Gun Club and were busy signing up members from around the world.

Jocelyn Francois from Lille in France is hitchhiking around America and watched from Cocoa Beach. She exclaimed: "C'est formidable! C'est magnifique!"

The Ayaldes and Pereas families travelled from Cali, Colombia. Raul Ayaldes said: "We came because it's the most important historical moment since the voyage of Columbus."

Around 10,000 people from Huntsville, Alabama – home of the Marshall Space Flight Center – came in a 250-strong caravan of vehicles, with many given paid leave to watch.

LIFT-OFF LATE ... BY A FRACTION

The Launch Director joked today that his team were late launching…by a fraction of a second.

At a post-launch press conference, Rocco Petrone described the mood among his Cape Kennedy team in the final moments of countdown.

He said: "You could feel the tension in the firing room. We all knew this was the big one. There was a certain amount of, let's say, static electricity in the air. Anyway, we were 724 milliseconds late at the launch of a mission that really started eight years ago! There was elation and a lot of pride."

Petrone, who handed on responsibility for the flight to Houston when Apollo 11 cleared the tower, added: "From the moment of truth here, less and less equipment has to work, so the numbers start to come on our side. But every step has its own dangers. I'm going to sweat every step until they're safely home. If anything happens and we fail, we push straight on with Apollo 12."

Apollo 11 faced several critical manoeuvres after launch, but Mission Control reported tonight that every stage of the Saturn V performed perfectly. Eight hours into the flight, the crew took off their spacesuits and when they settled down for the night, Armstrong reported they were all "as fit as a fiddle".

Doctor Wernher Von Braun (left) laughs with Doctor George Mueller (centre) and Lieutenant General Sam Phillips in the Kennedy Space Center control room after the successful launch.

In a press conference at Mission Control, Clifford Charlesworth – one of four Flight Directors taking it in turns to oversee the mission – said: "We've had an extremely good start. We've a few little 'funnies', but nothing we expect to have any impact on the flight."

Doctor Wernher Von Braun, who was at Cape Kennedy for the launch, said: "I'm very gratified at events so far, but I reserve my final comment until they're hale and healthy back on the ground."

The moment Apollo 11 achieved Earth orbit, it was given the number 4039 by the Air Force's Aerospace Defense Command, which keeps records of every man-made object launched into space. There are 1,743 currently in space, while another 2,000-plus have returned or burned up.

FLIGHT IS NO JOKE

The astronauts lived up to their reputation for being business-like in their chatter with Mission Control today.

TV viewers hoping for a repeat of the exclamations of delight and frustration heard during previous flights were left disappointed.

In the first minutes of Apollo 10's flight, Commander Gene Cernan exclaimed: "What a ride, babe! What a ride!"

His crewmate Tom Stafford had enthused: "Man, this is the greatest!"

In contrast, Neil Armstrong kept the chat to a minimum in the early stages and restricted his comments to reports on the spacecraft's flight status. One rare moment of levity came later today when the crew made an unscheduled fifteen-minute test TV transmission. During the telecast, which was recorded and screened to the public later, the view of the Earth was mostly of the Pacific Ocean. Buzz Aldrin asked Houston: "Hey, do you suppose you could turn the Earth a little so we could get more than just water?"

JAN'S BOAT TRIP

Neil Armstrong's wife and children dodged the media spotlight by watching the launch from a boat.

Jan Armstrong had been expected to be in the VIP seats at the space centre, but she and sons Ricky, 12, and Mark, 6, were on the Banana River five miles south.

Before flying back to Houston this afternoon, she said: "We were all very quiet and in awe of this spectacular we were watching." Ricky said: "We couldn't see the rocket right away. I was kind of worried. All of a sudden we could see it and it was beautiful."

The Armstrongs were among tens of thousands who watched from the water, with Brevard County officials estimating there were 5,000 craft on the Banana and Indian rivers. Many were charter boats laid on by aerospace companies for their staff.

BREAK A LEG, BUZZ!

Buzz Aldrin's wife has revealed she gave him a superstitious theatrical send-off.

Amateur actress Joan Aldrin, addressing the media outside her home after watching on TV, was asked about her final phone conversation with her husband last night. She said: "I told him 'Go break a leg!'" The phrase is used in theatrical circles as a way of saying good luck.

The astronauts' wives all have 'squawk boxes'; devices giving a constant link to mission radio communications. Joan said: "I'm going to be glued to that box every minute of the flight and I'm going to sleep right beside it. I'm excited, of course, and tense, of course."

Pat Collins – who, like Joan, watched with family and friends – said: "Wasn't it just wonderful? There were a few moments of tension. After that, lots of smiles. The children are excited, but they don't fully understand the significance. I'm glad the flight is under way. When we spoke last night, Mike sounded eager and all set to go. I think I felt about as ready as he did."

After launch, ten-year-old Kathleen Collins brought out a tray of coffees to the media, and told them about her favourite subjects at school.

NIXON'S ASTRO GUIDE

Richard Nixon got expert commentary from astronaut Frank Borman as they watched at the White House.

Borman, appointed by NASA as official mission liaison to the President, told Nixon it was a perfect lift-off and that crew-Mission Control communications were the clearest he'd heard during a launch.

Ronald Ziegler, White House Press Secretary, said Nixon "watched with interest" for thirty minutes before getting back to work and had checked on the flight's progress throughout the day.

'I FORGOT THE HUNGRY'

The man leading poverty protests at Cape Kennedy admitted today he was so overawed watching blast-off that he momentarily forgot all about the nation's hungry.

Reverend Ralph Abernathy, President of the Southern Christian Leadership Conference, and dozens of his protest group, watched from

VIP seats as guests of NASA. Abernathy said: "I really forgot the fact we have so many hungry people. I was one of the proudest Americans as I stood on this spot. This is holy ground and it will be more holy once we feed the hungry, care for the sick and provide for those who don't have houses. We must have a launching of a programme against poverty, hunger, racism and war. A launching that's just as effective and beautiful as the Moon shot."

Earlier, the protestors staged a march with wagons and mules beside Highway US-1. They were heckled by some motorists and there were tense moments when they were briefly stopped by police, who were concerned about the danger of them being on the main road. A County Sheriff's Office spokesman said: "They gave us quite a problem and caused a lot of close calls. We were thinking about locking them up, but we compromised and got them off the highway."

BROTHER'S SPACE PLEDGE

The brother of an astronaut killed in the Apollo 1 fire today vowed to follow him into space.

Ed White's brother, James, and the astronaut's widow, Pat, were in the VIP seats today and Air Force fighter pilot James, 27, said: "That crew's going to give America its most brilliant moment." He is due to leave for the Vietnam War soon, but asked if he'd like to be an astronaut, he said: "Yes sir! You bet!"

Pat White, whose husband was the first American to make a spacewalk, said: "A successful lunar landing would be the culmination of the hopes and dreams of many men."

HARRODS BUST-UP

Posh London store Harrods was the scene of an angry bust-up when TV sets in the electrical department were turned off moments before launch.

More than a hundred shoppers crowded around TVs in the luxury Knightsbridge store, but staff were ordered to pull the plug for safety reasons, sparking angry complaints. Stanley Rattenbury, Merchandise Manager, said: "We weren't being bloody-minded. The customers congregated to such an extent that it was dangerous."

BRAZIL, ARGENTINA BLACKOUT

Millions of South American TV viewers missed the launch.

Hundreds of millions in thirty-three countries on six continents watched, but Brazil and Argentina were blacked out as fears over a lack of available feeds – caused by a satellite failure last month – came true. Puerto Rico and the Virgin Islands were also hit.

Entel, Argentina's national communications company, complained that viewers in Europe and the USA had been given priority over South Americans.

AGNEW SPARKS MARS ROW

Vice President Spiro Agnew has sparked a heated debate over future space funding after using a TV interview to back a manned mission to Mars.

He told the nation: "It's my individual feeling that we should articulate a simple, ambitious, optimistic goal of a manned flight to Mars by the end of the century. Whether we say it or not, someone's going to do it."

Agnew, a member of a panel set up to recommend how the programme should proceed, said: "We're all advisers to the President. He'll make the ultimate decision. We have to consider that man stands on the threshold of great knowledge and to refuse to take the steps to acquire it may turn out to be the most debilitating thing as far as curing the ills of society is concerned."

Agnew also addressed NASA staff, drawing cheers and applause when he told them: "I bit the bullet for you today as far as Mars is concerned."

He admitted he may be "a voice in the wilderness" and today his statements received a mixed response. Democrats such as John F. Kennedy and Lyndon Johnson spearheaded the space race, but Republican Agnew's comments have exposed a divide among Democrats.

Michael Mansfield, Senate Majority Leader, said: "We have a lot of problems here on Earth that we must face up to and, when we settle those, we ought to consider future space ventures. The needs of the people on Earth – and especially in this country – should have priority."

Edward Kennedy, JFK's brother, is Senate Majority Whip and the Democrats' likely nomination for the 1972 Presidential election. He said: "After Apollo, the space programme ought to fit into our other national

Former President Lyndon Johnson (left) and Vice President Spiro Agnew watch from the VIP grandstand this morning as Apollo 11 soars into the distance.

priorities. We have important and demanding needs here on Earth that demand attention."

But another Democrat, House Whip Hale Boggs, argued: "We're only at the beginning. There have always been doubters. To stop now would be like Queen Isabella telling Columbus when he returned from his first trip 'That's enough'." Boggs was at the launch and said: "I had tears in my eyes and I'm not ashamed of it."

TAKING THE HIGH WAY

Neil Armstrong's mother wept as she watched Apollo 11 blast off, then joked: "I always told the children to aim high, but I never expected Neil to go quite *that* high!"

After watching at home in Ohio, Viola said the nerves she felt were similar to those before Neil – her first child – was born. She said: "I was so thankful for such a beautiful, beautiful launch."

Neil Armstrong waves to well-wishers this morning as the astronauts prepare to board a bus bound for the launchpad.

She spoke to her son on the phone last night and told him to "hurry home" so she could make him his favourite dish, apple dumplings. "He was happy, his usual self."

Stephen Armstrong said the launch was more exciting than watching Apollo 10 at the Cape "because our son was on top of it". He said: "I spent a better night last night than for four or five nights. I guess I just made up my mind that this was the day."

Buzz Aldrin's father, Edwin, told reporters outside his home in Brielle, New Jersey: "I was hoping everything would go smoothly, but you never know. I'm sort of a cool guy. I just get into a flap once in a while. I'll be excited when Buzz steps on the Moon. I do hope we get a good TV picture, especially of Buzz saluting the flag. Now, *that* will make some picture!"

PIGEON BEATS THE JAMS

A photographer beat the Cape Kennedy traffic and his newspaper's deadline today – thanks to a pigeon.

The *Daytona Beach Evening News* offices are only sixty miles away, but John Gontner realised he'd never get his film there by road. So, fifteen minutes after launch, racing pigeon Silver set off at 45mph with the film and, when the 14oz bird touched down seventy-eight minutes later, darkroom technicians immediately set about processing it. At 2.55pm, the first copies of the paper began rolling off the presses, with Gontner's shots on page one.

'MOON DAY' HOLIDAY

Richard Nixon has called for a national holiday to celebrate the lunar landing.

Although the President can't make a public holiday mandatory at such short notice, he has declared Monday a National Day of Participation. He hopes that all but the most essential public workers will be given the day off and that private companies will join in.

Nixon said what the astronauts were set to achieve would "stand through the centuries as one supreme in human experience," and that "the history they're making is not theirs alone, but everyone's".

Homer Ford, a Councilman in the astronauts' home city, Houston, welcomed the idea. He said: "Everybody will be watching until three or four o'clock in the morning, so no one will be working much on Monday anyway."

JFK 'WATCHING IN HEAVEN'

John F. Kennedy predicted he'd watch the lunar landing while sitting on a rocking chair in heaven, his brother-in-law claimed today.

Speaking at today's launch, Sargent Shriver – who is married to JFK's sister, Eunice – said: "I regret that he isn't alive to see his dream come true."

But Shriver, US Ambassador to France, said Kennedy had made a prophetic prediction in 1961. "He and members of the family were sitting in his office. He had half a smile on his face as he talked about the idea of Americans going to the Moon. Then he told us 'I firmly expect this commitment to be kept. And, if I die before it is, all of you should just remember that I'll be sitting up there in heaven in a rocking chair and I'll have a better view of it than anybody'."

NO LUNA UPDATE

Soviets were kept largely in the dark today over the race to the Moon.

No updates were given on the USSR's mystery probe, Luna 15, and there was no live coverage of Apollo 11. State TV and radio did, however, eventually break their week-long silence over the American mission, confirming the launch in later news bulletins.

A fleet of Soviet warships had a perfect view of the launch, having anchored seventy-five miles off the Florida coast. A US Department of Defense spokesman said: "They're there legally and Apollo is an open programme. We have nothing to hide."

MINER DIGS SPACE

English coal miner Bill Henshaw was allowed home early today for his own space mission.

Henshaw, 33, from Mansfield, Nottinghamshire, is one of only three UK amateur astronomers chosen by NASA to take part in its worldwide tracking operation.

The man known as 'Starry Eyed Bill' at Pleasley Pit, Nottingham, was given an early cut so he could have his £500 telescope ready for the first night of the mission. The British Astronomical Society member said: "I can hardly wait for the sunset. Many times, through my telescope, I've seen the area on the Moon where the astronauts hope to land. I shan't be able to see them in detail, but every night I'll be looking for anything unusual."

ASTRONAUT DIVORCED

NASA chalked up an unwanted landmark today – the first marriage of an active astronaut to end in divorce.

While the eyes of the world were on Cape Kennedy, Donn Eisele's wife, Harriet, was being granted a divorce in Houston. Her application to end their sixteen-year marriage on the grounds of mental cruelty was uncontested by the Apollo 7 astronaut. NASA has always been keen to present an image of its astronauts being happily married family men.

AEROSPACE NOSEDIVE

The Apollo 11 feelgood factor helped give America's stock market one of its best days in months.

The Dow-Jones Industrial Average registered its biggest one-day increase since April, but analysts don't expect the mission to have any long-term impact.

Aerospace industry stocks weren't among those to benefit from today's boost and experts believe there are tough times ahead for Apollo contractors. Michael Slutsky, Aerospace Analyst for Argus Research Corporation, said: "These companies have already built the hardware for the Apollo project. It's ironic that, in this moment of great achievement, the aerospace industry's stocks are in the doldrums."

CHAPTER 17

THURSDAY, 17 JULY
Launch Day Plus One

"It looks like the American people are backing it, which is more than I can say for what we're doing here in Vietnam."
American Soldier Jerald Drain

"Why should blacks rejoice when two white Americans land on the Moon when white America's money and technology haven't even reached the inner cities?"
Civil Rights Activist Marion Barry

Buzz Aldrin says the astronauts have been enjoying "out of this world" views of the Earth as they continue on their journey to the Moon.

PAST THE HALFWAY MARK

The astronauts took it easy today as they coasted past the halfway mark on their 238,857-mile ride to the Moon.

Shortly after crossing that imaginary line, they successfully carried out their one critical task; firing the main engine to fine-tune their course. Clifford Charlesworth, Flight Director, confirmed they were bang on course and that "the spacecraft is in good shape".

After sleeping soundly on their first night in space, the trio carried out basic mission duties and Buzz Aldrin joked: "Sure is nice here. I've been very busy so far. I've been cooking and sweeping and almost sewing and, well, you know, the usual housekeeping things."

The craft reached the halfway mark at 11.25am – twenty-five hours and fifty-three minutes into the flight – and had slowed to 3,689mph from the top speed of 24,200mph it reached yesterday when it surged out of low-Earth orbit. With the pull of Earth's gravity gradually slowing the craft, it will take twice as long to cover the second half of the journey.

Today's 2.9-second firing of the engine in the rear of the service module increased the craft's speed by just 14mph to ensure it passes within sixty-nine miles of the far side of the Moon on Saturday. Without the correction, they would have been 132 miles off course.

Michael Collins had a frustrating time trying to make star sightings using a sextant. He had trouble lining up the stars in the reticle – the device's crosshairs – but told Mission Control: "It's really a fantastic sight through that sextant. Just a beautiful sight. A minute ago, the reticle swept across the Mediterranean. You could see all of North Africa, absolutely clear, all of Portugal, Spain, southern France, Italy; all absolutely clear. But still no star!"

Aldrin spent some time observing the Earth through a telescope and said: "To coin an expression, the view is just out of this world. I've been having a ball floating around in here, back and forth, from one place to another."

SOVIETS RACE AHEAD

The mystery Soviet space probe entered lunar orbit today, raising the real possibility that America could lose the race to return Moon samples to Earth.

Jodrell Bank Observatory in England is the only facility outside the USA capable of tracking communications from Luna 15 and its Director,

Sir Bernard Lovell, said: "We have the fascinating situation of listening to the Apollo astronauts in one ear and signals from Luna 15 in the other.

"Tomorrow could be extremely interesting. We may find Luna 15 has touched down on the Moon. It's my opinion that an attempt will be made to land and bring back samples. I find it difficult to believe this is just another lunar orbiter. If the Russians intend to put Luna 15 in orbit and just leave it there, the whole operation is incomprehensible. It simply doesn't make sense."

The news that Luna 15 had achieved lunar orbit was relayed to Apollo 11 by Bruce McCandless, duty Capsule Communicator (Capcom) in Houston. The astronauts didn't comment on the news.

Meanwhile, a fleet of Soviet warships that were anchored off the Florida coast for the Apollo launch are now on their way to Cuba to join celebrations of the tenth anniversary of the revolution there.

SO QUIET IN SPACE

Space chiefs have hit back at critics who have branded the astronauts boring.

Some media commentators have highlighted a lack of banter between the crew and Mission Control, but Clifford Charlesworth, Flight Director, was clearly irritated when the issue was raised during a press briefing today. Asked whether he'd get annoyed at the criticism if he were one of the crew, he said: "I would and I *do*."

When it was suggested NASA should encourage them to open up, he said: "This particular crew is just not talkative. I think that's pretty clear. They're just a quiet crew. I certainly wouldn't want to backseat drive for them."

Charlesworth raised a laugh when asked what he thought Neil Armstrong might say when he and Buzz Aldrin land on the Moon. "I expect he'll call Houston and say he has landed."

Communications were mainly business-like today, although Capcom Bruce McCandless did give the crew a news bulletin including concerns being raised in the UK's House of Lords that an American mini-submarine exploring Scotland's Loch Ness might harm the Loch Ness Monster.

Later, Michael Collins initiated a session of running on the spot in weightlessness. He asked: "Hey! You got any medics there watching heart rates? We're all running up here. You wouldn't believe it!" Houston told Collins his heartbeat had risen to 96bpm.

SPACE TV SHOW

The astronauts put on a TV show for the world tonight, oblivious to the debate over their perceived lack of personality.

They appeared comfortable and relaxed in a broadcast lasting thirty-five minutes. The first two-thirds of the show concentrated on images of the Earth, with the USA, Canada, Mexico and Central America all clearly visible.

Then Neil Armstrong announced: "Okay, world! Hold on to your hat! I'm going to turn you upside down." Still clutching the hand-held camera, he rolled over several times before conceding: "I'm making myself seasick doing this. I'm going to put you back right-side up."

Michael Collins gave viewers a guided tour of the command module cabin and the pictures were so sharp that viewers could even read the digits on instrument panels. Switching his attention to the pantry, he said: "We've got all kinds of good stuff: coffee, various breakfast items, bacon in little, small bites." Holding up a plastic package, he added: "Would you believe you're looking at chicken stew? All you have to do is add three ounces of hot water and blend for five or ten minutes. Beautiful chicken stew. The food so far has been very good."

AND NOW, THE WEATHER

Neil Armstrong turned weatherman today to give a perfect forecast of conditions outside the Manned Spacecraft Center.

Armstrong radioed to say: "Houston, we're just looking at you out of the window here. Looks like there's a circulation of clouds that just moved east of Houston over the Gulf and forward area. Did that have any rain, do you know, this morning?"

The control room has no windows, but after a brief delay Capcom Bruce McCandless confirmed: "Roger. Report from outside says it's raining. Looks like you got a pretty good eye for the weather." Armstrong responded: "Looks like it ought to clear up pretty soon. The edge of the cloud pattern has almost reached you."

His forecast was spot on. While they were speaking, a thunderstorm was passing over the space centre. Minutes later, as Armstrong predicted, the rain stopped.

DEAD SPACEMEN HONOURED

Commemorative medals honouring five dead space pioneers will be left on the Moon, it was announced today.

Richard Nixon revealed the special tribute will be made to Apollo 1 victims Gus Grissom, Ed White and Roger Chaffee, as well as Soviet cosmonauts Yuri Gagarin and Vladimir Komarov.

Gagarin, the first man in space, died in a plane crash last year. Komarov was killed two years ago when his Soyuz 1 craft crashed on landing. Their widows gave medals to Frank Borman during his visit to the USSR and asked if they could be taken to the Moon.

Nixon stressed there was "no national boundary to courage," and said it was "fitting that the first lunar explorers carry with them some recognition of the sacrifice made by other space pioneers who helped blaze the trail".

Medals honouring the Apollo 1 crew of (left to right) Gus Grissom, Ed White and Roger Chaffee are to be left on the Moon.

THURSDAY, 17 JULY

SPACE FEVER SWEEPS GLOBE

Excitement over the lunar mission was sweeping the world today.

Provisional TV viewing figures released by ABC estimated that 528 million people watched the launch and the Moonwalk is expected to be the most-watched event in history, with an audience of more than a billion.

Across the world, special preparations are being made for Sunday's historic events. Giant TV screens are being erected at numerous sites, including Central Park in New York, Rio de Janeiro's Museum of Modern Art and Nathan Philips Square outside Toronto City Hall in Canada. Footage will be replayed throughout Monday for visitors to Disneyland, California.

An English pub called 'The Man on the Moon' was today given permission to stay open late so customers can watch TV coverage of the landing. Touchdown will coincide with the UK Sunday closing time of 9.30pm, so magistrates in Norwich, Norfolk, have granted the landlord a ninety-minute extension.

In Pakistan, a baby born at a Red Cross maternity home shortly after launch has been named Apollo. Nurses who delivered the 8lb 11oz boy suggested the name and the parents agreed.

Indigenous Canadians living in remote arctic areas are following the mission via specially-translated radio reports. Broadcaster CBC has six shortwave stations serving northern outposts and is translating coverage into languages such as Inuit. Spokeswoman Lucy Mackay said: "The Eskimos are especially interested. They don't have nebulous words like liberty or freedom in their language, but the Sun, stars and planets have a strong sway over their way of life."

American retailing giant Sears Roebuck revealed today that sales of telescopes this year are already double what they were in the previous twelve months. A spokesman said the company had given salesmen strict instructions to stress to potential buyers that they won't be able to see the astronauts on the lunar surface.

The first woman to fly into space, Soviet cosmonaut Valentina Tereshkova, today wished Apollo 11 success and predicted women will be on future Moon missions. Tereshkova, speaking during a visit to Poland, said she believed her forty-eight Earth orbits on Vostok 6 in 1963 proved that a woman's body could withstand the demands of space just as well as a man's.

Today's newspapers were dominated by the mission and the coverage was almost all positive. There were exceptions, though. A satirical cartoon in Britain's *Daily Mail* showed two starving children in famine-stricken Biafra gazing up at the Moon and saying: "Maybe they'll discover it's made of cheese and bring some back for us to eat."

BIG NOISE IN FLORIDA

People living more than a hundred miles from Cape Kennedy could hear the roar of the Saturn V blasting off, it emerged today.

Residents in Anthony, 140 miles inland, Moss Bluff, 118 miles from the Cape, and Ocala, 132 miles away, all reported hearing the launch. In Anthony, locals even felt the ground vibrate.

Moss Bluff resident Margarette Hornbeak said people clearly saw the first stage of the rocket separate and a spurt of flames as the second stage fired.

Meanwhile, the crew of German ship *Vegesak* reported that numerous pieces of the rocket fell into the sea around them, 375 miles east-north-east of the launch site. No damage was caused.

HOLIDAY CONFUSION

America was in a state of confusion today over Richard Nixon's surprise national holiday proclamation.

So far, thirty-one State Governors have confirmed they will declare a public holiday on Monday, but even in those states some Mayors have indicated it will be a normal working day for city employees.

Some major private employers say they will close, but it's likely to be business as usual for the majority. With department stores in Philadelphia, Washington and Boston all saying they'll be open, others across the country will probably follow suit and most of the nation's 4,700 banks will be open, despite an appeal by William Camp, Controller of Currency.

Even Apollo contractor Boeing won't be shutting on Monday. A spokesman said: "Our enthusiasm for this achievement is great. However, our other commitments are so pressing that it's essential we remain on the job."

CALL TO BOYCOTT CELEBRATIONS

African-Americans have been urged to boycott Richard Nixon's 'Moon day'.

Marion Barry – co-founder of Pride Inc. Operations, which provides work for unemployed African-Americans – called a press conference to speak out against the national holiday.

He said: "The feat of men landing on the Moon – invading the Moon – is only a quasi-accomplishment, in that black people have no relevance to this whole invasion and desire no part of it. Why should blacks rejoice when two white Americans land on the Moon, when white America's money and technology haven't even reached the inner cities?

"Why should blacks marvel at white America's ability to build a complex spaceship when white America's technology hasn't developed a way to build low-cost housing for blacks and poor whites? We're calling upon the entire black community not to observe Monday as a holiday, but to keep working."

Barry's call came as America was rocked by race riots in Youngstown, Ohio, after an African-American woman complained she was kicked and racially abused at a dairy store by the owner.

SECRET CARGO

The astronauts' wives are staying tight-lipped over personal mementoes their husbands have taken on the mission.

NASA has confirmed the crew were given permission to take small items that have a special meaning to themselves and their wives, but that the astronauts had asked for the details to be kept private.

Joan Aldrin said: "Yes, Buzz took something very special for me, but I'd rather not say what." Of how she was feeling after launch, she explained: "I'm much more relaxed now that the mission is under way. I slept well. You just have to stay with it and hang loose."

Pat Collins wouldn't be drawn on what her husband had taken and Jan Armstrong said simply: "That's private."

Jan told a press conference outside her home that she rated the mission's chances of success at "100 per cent". When reporters responded with "Are you really *that* confident?" and "Are you a little fearful?", she replied "Yes, I am," and "No, sir, I'm not." She added: "The launch was a tremendous sight. I was just thrilled. For me, though, there won't be any celebrating until they splash down."

On how the couple would cope with Armstrong's fame as the first man to walk on the Moon, she said: "I'm not worried about being a celebrity. If Neil has to spend a great deal of time with other things, then we'll try to make the most of the time we do have together."

GOOD LUCK CHARMS

Neil Armstrong's parents are both wearing good luck charms given to them by their son.

Stephen Armstrong, revealing that the astronaut gave them pin badges from his Gemini 8 flight, pledged: "I'll wear this all the durn time he's gone."

Viola Armstrong said: "We're looking forward to after this flight when we hope Neil will be able to come home more often. He doesn't get home nearly enough."

Buzz Aldrin's dad arranged a police guard outside his New Jersey home to keep the press at a distance. The 73-year-old retired Air Force Colonel did emerge for a spot of verbal jousting, though. Asked what he told his son when they spoke on the eve of the launch, he said: "What did I say? I said 'God bless you and have a safe trip'. What did you expect me to say?"

THE PRIDE OF OHIO

Neil Armstrong is proving to be a big cheese in his home state.

Plans were revealed today for a Neil Armstrong Museum in Ohio, while a dairy company in his home town is booming thanks to Moon Cheese.

Jim Rhodes, State Governor, said details for the museum were already being drawn up. "This should be a place to keep all of the personal mementoes of Neil Armstrong on his historic flight to the Moon and a place for a detailed, permanent record of man's first landing on territory in outer space."

Rhodes added: "I would also expect officials to consider the naming of schools, libraries, streets and other public facilities after our own Neil Armstrong."

In Wapakoneta, the Fisher Cheese Company is the town's biggest employer and is struggling to keep pace with demand for Moon Cheese, a yellow cheddar it started producing after Apollo 8's lunar mission. The packaging features a picture of the Moon, a space capsule and an astronaut, plus the slogan 'Unofficial sample of the Moon'.

Fisher has expanded to a seven-day-a-week operation and Carl Kaplanoff, Director of Public Relations, said today: "I don't know how we're going to handle the business. We're getting orders from all over. We sent 5,000lb to Florida and they still want more."

SOLDIERS TUNE IN

American soldiers are following the Moon shot amid the turmoil of the Vietnam War.

The Armed Forces Network is keeping the 500,000-plus troops up to date with radio reports and daily TV screenings of mission highlights.

With the US death toll rising to 37,167 after today's confirmation that another 148 were killed in combat last week, there is a marked contrast at home between the optimism over Apollo 11 and the bitter opposition to the war.

Private Jerald Drain, 24, from Quitman, Georgia, said of the Moon shot: "It looks like we've got worldwide support and the American people, especially, are backing it, which is more than I can say for what we're doing here."

Master Sergeant Edward Conner, from Hawthorne, New Jersey, said: "The astronauts have a hell of a lot of guts. I wouldn't want to try it myself."

SHARK ATTACK SCARE

The astronauts could face another deadly threat if they survive the Moon mission – a shark attack.

Rehearsals for next week's splashdown in the Pacific are being blighted not only by bad weather but by schools of sharks. Divers from the USS *Hornet* recovery ship have several times had to scramble back on to rubber rafts to escape the predators.

Today, sharks chewed on equipment and Lieutenant Wes Chesser, team leader for the divers, said: "There are sharks out there, swimming all over the place."

Seaman Jack Wolfram, 20, will be the first diver in the water after splashdown, and he said: "The sharks have come right out of the water. The sharks watch us and we watch them. On the day, I'll be alone in the water for between three and five minutes. I'll be looking for sharks and giving a signal whether there are any in the water."

SPACE AGENCY v AD AGENCIES

NASA today hit out at advertisers for cashing in on the Moon mission and admitted it's virtually powerless to stop them.

Linking products to space exploration has become a favourite tactic for the advertising agencies of New York's Madison Avenue. Julian Scheer, NASA Assistant Administrator for Public Affairs, said: "The increasing frequency of such advertising is beginning to give it a bad flavour. The American taxpayer is footing the bill for the space effort and there shouldn't be exploitation."

Everyone from department stores to plumbers are jumping on the bandwagon and Scheer said: "We don't encourage it, but in our system of free enterprise we're powerless to do anything about it. That doesn't mean we don't have a very uneasy feeling about it."

For companies which *are* involved in the space effort, NASA forbids adverts to say it has 'approved' or 'endorsed' a product, only that it has 'selected' it.

New York-based Duro makes the marker pens being used on Apollo 11 and is advertising under the slogan 'The marker that went to the Moon'. Robert Perlumutter, Duro Vice President, said: "We feel this will add to our prestige. We're just trying to get the point across that our product must be superior to be used by NASA."

Some companies, conscious that adverts could backfire in the event of a lunar disaster, are waiting until the astronauts are safely home. Omega will launch a campaign after splashdown promoting the $195 watches the astronauts are wearing. The ad says: 'Omega – the first watch worn on the Moon salutes the men who took it there.'

Hasselblad will promote its cameras with the line: 'This is the Hasselblad that took the pictures of the Moon … it looks very much like an Earth Hasselblad.' A spokesman said: "We're just waking up to the unlimited possibilities of exploiting our cameras' use in space."

Lilly & Company is choosing *not* to exploit its Apollo connections. Its sleeping pill Seconal is in the astronauts' medical kits, but a spokesman said: "We don't advertise our connection with Apollo. Doctors have been prescribing Seconal for a long time. Any advertising would be pure Madison Avenue."

SURGE IN SPACE BOOKINGS

Airlines say the excitement over Apollo 11 has sparked a surge in reservations for proposed commercial spaceflights.

Najeeb Halaby, Pan Am Chairman, revealed its waiting list stands at 16,700 following a rush of applications from around the world since the launch.

Tex Tyrell, a 39-year-old auctioneer from Darwin, Australia, phoned Pan Am yesterday to reserve seats for himself and wife Lola. He said: "As soon as we saw the lift-off we said 'This is it'. Just think, 102 miles out in twelve minutes. It shouldn't be more than three or four years before they can take passengers into space. They can make computers today that can't make a mistake. It's only humans that make mistakes."

A spokesman at Pan Am's reservations centre said: "Naturally there's no estimate of the cost or when it'll be. I figure it won't be until after I'm long gone. There won't be any details of the flight available, I'd say, for ten or twenty years.

"All we can do is put people on the list and, when the time comes, start calling to say 'Are you going to chicken out or are you going with us?' I love talking to passengers about Moon trips, but I just don't think we're going to be doing it soon."

Air Canada has had sixty-two reservations. A spokesman said: "The flight will be traditionally high quality even though coffee, tea and milk may have to come from a tube."

CHAPTER 18

FRIDAY, 18 JULY
Launch Day Plus Two

"Everything is getting smaller and smaller."
        Apollo 11 Lunar Module Pilot Buzz Aldrin

"The tourists are always asking for Walter Cronkite and are disappointed he's not here. He's the housewife's delight; more popular than the astronauts."
        Apollo Contractors' Marketing Fair Hostess Mary Miller

An image taken from tonight's live telecast shows Buzz Aldrin carrying out checks inside the lunar module.

THE GREAT ENTERTAINERS

The astronauts made a mockery of their 'space bores' tag tonight with a spectacular live TV show from 200,000 miles into space.

An estimated audience of fifty million were transfixed as Neil Armstrong and Buzz Aldrin floated into the lunar module to inspect for damage; a unique opportunity for viewers to see astronauts carry out critical work on board a spacecraft. The fragile Eagle was found to have come through launch unscathed.

The captivating transmission lasted one hour and thirty-six minutes – more than double the previous longest – and Capcom Charlie Duke thanked them for "one of the greatest shows we've ever seen".

The inspection began when Armstrong removed a docking device, allowing access to the lunar module. He reported: "Mike must have done a smooth job in docking. There isn't a dent or a mark."

With the hatch open, Aldrin wriggled into the tunnel between the modules to check for damage. There was relief as he confirmed: "Everything looks good down there."

The astronauts, appearing more relaxed than on the first two days of the mission, chatted and joked throughout. As he floated inside Eagle, Aldrin said of restraining straps he was attached to: "These are doing a good job of pulling my pants down!"

Michael Collins, who remained in the command module, introduced himself by saying: "Hello, Earthlings." When Duke asked if Collins would be going into the lunar module, Armstrong said: "He hasn't come up with the price of a ticket."

Former Eagle Scout Armstrong had one special message. He said: "I'd like to say hello to all my fellow Scouts and Scouters at Farragut State Park in Idaho. They're having a national jamboree there this weekend and Apollo 11 would like to send them best wishes."

The quality of the images was unprecedented and Duke told the crew: "We keep marvelling about the colour and clarity of the picture. It's just perfect." When Duke commented he could even see specks of dust floating about, Aldrin said: "Yeah, I'm choking on one every so often!"

The spacecraft travelled 2,300 miles during the show, which closed with pictures of the Earth 203,000 miles away.

'RIGHT ON THE BOOK'

Space agency bosses were in confident mood tonight as they geared up for the most critical seventy-two hours of the mission.

With Apollo 11 due to enter lunar orbit tomorrow, Glynn Lunney, Flight Director, said the craft's course was "right on the book". He said: "It's going as well as any mission we've ever had, that's for sure."

The astronauts had another easy day, carrying out chores including charging batteries, dumping waste water and checking fuel and oxygen reserves. They settled down for the night knowing another important landmark would be achieved while they slept; the point where the Moon's gravitational pull becomes stronger than that of the Earth and starts to increase their speed. The craft had slowed to 2,400mph earlier today, but will be travelling at 5,700mph tomorrow afternoon.

The crew had enjoyed a lie-in this morning, sleeping soundly for around nine hours. Capcom Bruce McCandless woke them after taking over microphone duties from Ron Evans, whose night-time stint is known as the 'Black Shift'. McCandless joked: "The Black Team is complaining they didn't get a chance to make any transmissions. Ron Evans is getting to be known as 'The Silent Capcom'." Michael Collins got a laugh when he replied: "That's the best kind, Bruce!"

At that point, Apollo 11 was 73,000 miles from the Moon. Buzz Aldrin, looking at Earth through a window, observed: "Everything's getting smaller and smaller."

SOVIETS' LUNA PLEDGE

The Soviet Union today cleared America's path to the Moon by promising its lunar probe won't get in the way.

In an unprecedented step, a Soviet space official shared details of Luna 15's flight plan and said its trajectory wouldn't interfere with Apollo 11. NASA asked Frank Borman to approach contacts he made on his USSR tour and Doctor Mstislav Keldysh, President of the Soviet Academy of Sciences, cabled the details.

Julian Scheer, NASA Assistant Administrator for Public Affairs, said: "I can recall no other direct communication like this." Doctor Chris Kraft, Director of Flight Operations, said: "They gave us what we needed to know. I trust what they gave us explicitly."

Borman explained: "I called Doctor Keldysh and said 'There's some concern about the trajectory of Luna 15'. I asked him to check. I'm very pleased with the response. I hope it's indicative of more fruitful co-operation in the future."

It is thought Luna 15 will attempt to land and retrieve lunar samples and, although it remained in orbit today, Sir Bernard Lovell, Director of the Jodrell Bank Observatory in England, said he was convinced "something else is going to happen".

Doctor Elbert King, Curator of the Houston laboratory where Apollo 11's Moon samples will be studied, said an automated lander randomly scooping up material couldn't be compared with two highly-trained astronauts selecting samples. King said: "It would be like grabbing in the blind."

LANDING RISKS 'ARE LOW'

A top NASA official today insisted the mission is less risky than pioneering flights made by daredevil early aviators.

Doctor Chris Kraft, Director of Flight Operations, told a press conference he has total confidence in the lunar module. He claimed that

Doctor Chris Kraft, pictured here (left) with Doctor Robert Gilruth, believes the chances of the Apollo 11 mission failing are "low".

Charles Lindbergh, who flew solo across the Atlantic in 1927, and Wiley Post, who made a light aircraft flight around the world in 1933, were at greater risk than the astronauts.

Kraft said: "Engineering talent has been concentrated on reliability. I suspect it's as safe as a transport plane flight, certainly if you don't include the landings and take-offs. The probability of failure is estimated as low. A phenomenal amount of testing is involved to prove it all works. This helps explain why what we build is reliable. Lindbergh and Post took more risks in some ways."

Asked about concerns over Eagle landing at an angle, Kraft said: "Probably we can launch if it slopes as much as thirty-five degrees. If it landed in a crater, it might slide down but still be pretty safe."

NIXON TV STUNT 'TASTELESS'

Richard Nixon has been accused of a cheap publicity stunt after the White House revealed he will chat with the astronauts during the Moonwalk.

The President's historic call to the Moon will be screened live on TV around the world and today *The New York Times* criticised the President for using the mission for personal political gain. In a scathing editorial, the newspaper said: "Of the three Presidents who have occupied the White House since the initiation of Project Apollo, President Nixon has had least responsibility for that massive programme."

The newspaper argued that "apart from objections on grounds of taste", the TV broadcast shouldn't be allowed to eat into the limited time the astronauts have to explore the surface. "His attempt to share the stage with the three brave men on Apollo 11 when they attain the Moon appears to us rather unseemly. Such an intrusion looks suspiciously like a publicity stunt of the type Nikita Khrushchev used to indulge in. It strikes us as unworthy of the President of the United States."

Referring to signatures on a commemorative plaque to be left on the surface, *The Chicago Daily News* said: "President Nixon will be associated with the Moon flight, originated by President Kennedy, almost every step of the way. And it will be Nixon's name – not Kennedy's – that will stay on the Moon." *The New York Post* claimed: "Nixon sees political benefits in the historic feat."

Ronald Ziegler, White House Press Secretary, insisted the Moon chat was suggested by NASA and that "the President will speak on behalf of the American people".

WIVES' BBQ GET-TOGETHER

The Apollo 11 wives forgot their Moon worries today by lapping up some sun at a poolside barbecue.

Joan Aldrin hosted the get-together at her home in Nassau Bay and Jan Armstrong and Pat Collins attended with relatives. The guests made the most of the soaring temperatures by taking a swim in the backyard pool.

As Pat left she shouted to the waiting media: "Plenty of women's talk? You better believe it!" Commenting on the previous night's live telecast, she described it as "pretty wonderful", while Jan said: "It was great. Even better than I expected."

$1 MILLION 'LITTER BUGS'

Neil Armstrong and Buzz Aldrin will become the most expensive 'litter bugs' in history...by leaving equipment worth more than $1 million on the Moon.

A $250,000 TV camera and two emergency backpacks worth $300,000 each will be among the 'rubbish' they abandon, along with tools used for collecting samples. Smaller items, such as used food bags, will also be left behind and a NASA spokesman said: "They won't have any waste baskets or garbage cans, so they'll just have to leave it all when they're done with it."

CONTRACTORS FEEL THE HEAT

Dozens of Apollo contractors are camped out at a Houston hotel, with one eye on drumming up new business and the other on the success of the mission.

Representatives of more than thirty companies involved in the programme are taking part in a marketing fair at the Sheraton Kings Inn. As well as trumpeting their part in the Moon shot by giving out product samples, brochures and press releases, they're also acutely aware of the negative impact a failed mission could have.

North American Rockwell is Apollo's biggest contractor and Earl Blount, Space Division Public Relations Chief, admitted: "It's awfully tense. You stop worrying if your kid won the Little League ball game or whether your lawn sprinkler system is working. This is it!"

A stand set up by Ohio-based frozen food firm Stouffer's is proving to be one of the most popular at the fair. Visitors can sample meals the astronauts will eat during post-flight quarantine. Iris Lochner, of Stouffer's, said: "We have everything – pot roast, Swiss steak, desserts – and we've given out more than 2,500 samples so far. We've got many freeloaders, of course, who eat here three times a day. They love it, so what can we do?"

Mary Miller, hostess for the event, said: "The tourists who look in are always asking for Walter Cronkite and are disappointed he's not here. He's the housewife's delight; more popular than the astronauts."

1969: A TV ODYSSEY

A hi-tech device inspired by a mutinous computer in *2001: A Space Odyssey* is taking a TV network's coverage into the space age.

CBS has built HAL – named after the movie's HAL 9000 – to co-present its shows by displaying graphics and footage on four giant screens. It can interact with anchorman Walter Cronkite by displaying text on its screens.

In the movie, HAL 9000 malfunctions and stages a mutiny. Robert Wussler, CBS News Executive Producer, said: "HAL will play a key role, providing information and answering Walter Cronkite's questions. And we don't expect *this* HAL to try any takeovers!"

BBC SLAMS 'MOON PARTY'

Britain's public service broadcaster has accused its commercial TV rival of trivialising the mission.

ITV is spending £200,000 on its most expensive ever broadcast, which will feature performances from pop stars Cliff Richard, Englebert Humperdinck, Cilla Black and Mary Hopkin, plus comedy sketches featuring Ken Dodd, Eric Sykes and Hattie Jacques.

Ronnie Noble, who is producing the BBC coverage, said: "ITV are treating the whole thing as a slap-up Moon party. We'll be specialising in hard news. There will be no party atmosphere about *our* coverage."

NO OVERTIME PAY IN SPACE

The astronauts won't be paid overtime for working on Monday, despite Richard Nixon declaring it a national holiday.

Millions of public and private employees will get an extra day's paid leave, while in states and cities which have declared a holiday, essential workers will get overtime pay.

But Neil Armstrong's job falls within a bracket of the civil service which doesn't qualify for overtime. The same applies to serving Air Force officers Michael Collins and Buzz Aldrin.

The crew will, however, get more in their next pay packets because both the civil service and armed forces have agreed wage rises. Armstrong's salary will go up from $27,401 to $30,054, Aldrin's will increase from $18,623 to $20,607 and Collins' from $17,148 to $18,648.

Nixon's holiday idea continued to receive a mixed response today. The Post Office has announced that most of its 750,000 staff will get Monday off.

Kentucky is the latest state to reject Nixon's proposal. Wendell Ford, Lieutenant Governor, said: "Most Kentucky state workers will, I'm sure, be watching their TV sets at 2.15am on Monday and all of us may be a little sleepy when we get to work that morning. But I'd ask them to rededicate their efforts, do a good job and meet the down-to-earth needs of people here in Kentucky."

ANCIENT MOONWALKERS

The USA's second biggest group of Native Americans will observe Monday's national holiday – to honour the *second* pair of humans to walk on the Moon.

Members of the Navajo Nation – which spans parts of Arizona, Utah and New Mexico – believe they are descendants of ancient space travellers. Raymond Nakai, Tribal Council Chairman, confirmed today that Monday will be a "day of observance". However, his people don't consider the landing to be a first, as a Navajo legend tells of two boys who stopped off on the Moon on the way to visit their father, the Sun.

Ray Erwin, Mayor of Gallup, Arizona, has sent a telegram to Richard Nixon asking him to tell the astronauts to "check the surface of the Moon for moccasin prints".

SPACE LAWS REVIEW

The UN wants stricter space laws to avoid disputes on Earth spilling into the galaxy.

Nations including the USA and USSR are already bound by the 1967 'Outer Space Treaty', which forbids territorial claims in space. There has been controversy over the planned planting of the Stars & Stripes on the lunar surface, but the USA has stressed it won't be claiming the Moon as a colony. General Martin Menter, Chairman of the American Bar Association's Committee on Aerospace Law, said: "For the first time in history, planting a flag on unexplored territory means nothing legally."

However, the UN Outer Space Committee is discussing how it should tighten up laws for future exploration. Yvon Beaulne, Canadian Ambassador to the UN, said: "There are other legal questions. How is it possible to prevent the use of outer space for unwarranted intrusion into countries' defence systems? How can you prevent spying in outer space?

"I'm sure that a few years ago these questions would have appeared to be far-out, science fiction ideas, but these are questions of great importance which member states are now examining with great sincerity."

NEW KENNEDY TRAGEDY

The Kennedy family was rocked by another tragedy today on the eve of John F. Kennedy's lunar landing dream becoming a reality.

JFK was shot dead in 1963 and his brother, Robert, was also assassinated last year. Today, another of the brothers, Edward, was involved in a car accident in which Robert's former secretary died.

Edward managed to swim to safety when the car he was driving plunged off a bridge and into a pond on Chappaquiddick Island, Massachusetts, but passenger Mary Jo Kopechne, 29, drowned.

CHAPTER 19

SATURDAY, 19 JULY
Launch Day Plus Three

"The view of the Moon is really spectacular. It's a view worth the price of the trip."

Apollo 11 Commander Neil Armstrong

"I'm green with envy. I'd give anything to be up there right now. Who wouldn't?"

Former Astronaut John Glenn

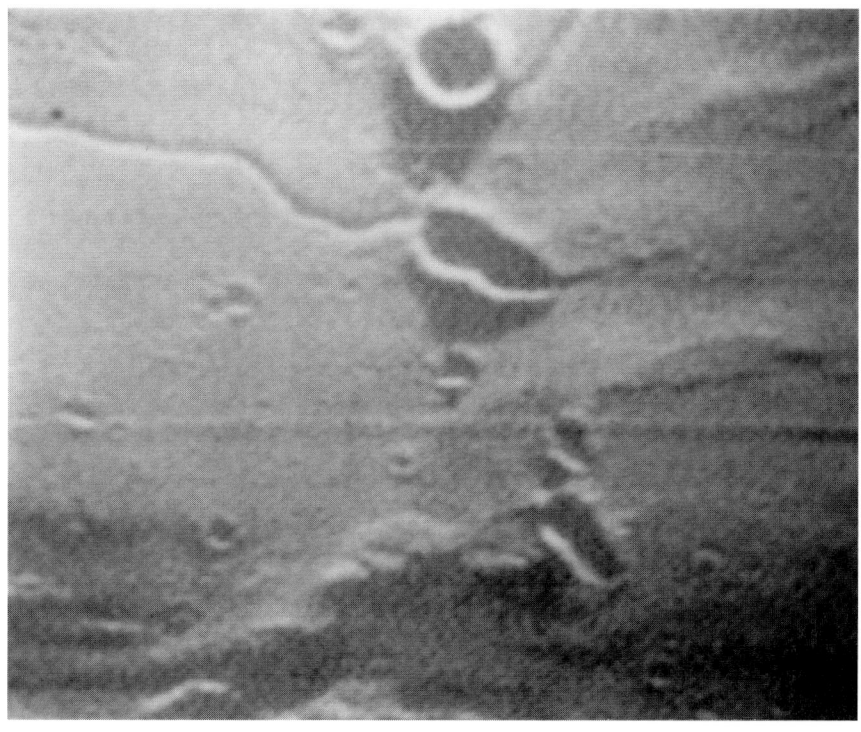

A view of the rugged lunar surface seen by TV viewers during today's live telecast.

APOLLO 11 IN LUNAR ORBIT

Apollo 11 was just twenty-four hours from making history tonight as it swooped around the Moon ready for tomorrow's landing.

In the most critical moment since launch day, command module Columbia fired its engine to achieve lunar orbit. As Apollo 11 was on the far side of the Moon at the time, Mission Control endured a tense thirty-five minutes of radio silence before contact was re-established.

Had the engine failed to fire, Columbia would have flown once around the Moon at 5,700mph and then returned to Earth. Had it fired for too long, the craft would have crashed. There was wild applause at Mission Control when Neil Armstrong reported lunar orbit insertion (LOI) had been successful.

Today's drama began at 1.12pm when, with the craft about to go behind the Moon, Capcom Bruce McCandless told the crew: "All systems are looking good going around the corner. See you on the other side."

NASA Public Relations Officer, Jack Riley, provided a live commentary for TV and radio. He described how many current and former astronauts were at Mission Control to follow events.

Had Apollo 11's engine not fired, Columbia would have re-emerged nine minutes earlier than expected. When that moment passed, Riley reported: "We're past the no-burn acquisition time now and we've received no signal. It's very quiet here in the control room." As loss of signal passed thirty minutes, he commented: "There are a few conversations taking place here, but not very many. Most people are sitting quietly, watching and listening, not talking."

After thirty-four minutes of radio silence, NASA's tracking station in Madrid, Spain, picked up data signals and McCandless twice radioed the astronauts, asking: "Apollo 11. This is Houston. Do you read?" Finally, after another minute's silence, the faint voice of Armstrong was heard saying: "Houston, Apollo 11. Over."

Once clear radio contact was re-established, Armstrong said the 20,500lb-thrust engine had performed "like perfect". The burn slowed Columbia from 5,700mph to 3,700mph, allowing it to surrender to lunar gravity and go into orbit at an altitude ranging between 70 and 195 miles.

After two circuits, the engine fired again for seventeen seconds to achieve the more circular orbit required for landing, with the craft's altitude now sixty-two to seventy-five miles. Later, Buzz Aldrin crawled into the lunar module and carried out a two-hour inspection, including turning on Eagle's systems for the first time since launch. He reported: "Everything looks super. We're ready to go."

Clifford Charlesworth, Flight Director, said tonight that there was nothing to stop the landing attempt going ahead. Charlesworth said a leak of gaseous nitrogen when Columbia fired its engine to achieve lunar orbit was nothing to worry about.

MOON CLOSE-UPS ON TV

TV viewers were treated to a close-up of the Moon's rugged surface tonight as the astronauts made a live telecast from lunar orbit.

The astronauts pointed out landmarks the lunar module will fly over during tomorrow's descent, but because the Sun hadn't risen over the Sea of Tranquility, the astronauts and viewers were unable to get a clear view of the landing site.

Within the space of six hours today, the crew went from catching their first view of the Moon from a distant 15,700 miles to seeing it from less than a hundred miles. Armstrong said: "The view of the Moon that we've been having recently is really spectacular. It fills about three-quarters of the hatch window. Part of it's in complete shadow and part in earthshine. It's a view worth the price of the trip."

Once in lunar orbit, Armstrong said he could see craters and "good-sized blocks" of rock. He described the landscape as "ashen grey" and, as they passed over the Sea of Fertility, he joked: "It doesn't look very fertile to me!"

He added: "Maps and pictures have given us a very good preview of what to look at. But, like the difference between watching a real football game and watching one on TV, there's no substitute for actually being there."

LIFE-OR-DEATH TEST

Neil Armstrong will have to live up to his reputation for excelling under pressure when he attempts to land on the Moon tomorrow.

His and Buzz Aldrin's lives will depend on snap decisions made by the Mission Commander. For much of the descent, the pair won't be able to see the surface, as the lunar module will be facing the wrong way until manoeuvring rockets fire to tip Eagle into an upright position.

Floyd Bennett, chief of NASA's Lunar Landing Analysis Branch, said: "With two minutes to go, the astronauts will have a very good view of the surface and it's anticipated that Commander Armstrong will take over and fly manual."

Armstrong can throttle the descent engine to hover above the surface for about two minutes if necessary as he searches for a safe landing spot. Bennett said the last possible moment to abort and rendezvous with the command module would be "about twenty seconds before he runs out of gas". Bennett said: "He's got to make up his mind and either land or get out of there."

The drama will begin while Apollo 11 is making its thirteenth orbit, with Eagle separating from the command module. Michael Collins will make a visual inspection for any exterior damage.

Ninety minutes after separation, Eagle's descent engine will fire it into a lower orbit. Then, fifty-six minutes later, the on-board computer will display '99' – a signal asking the astronauts whether they want to begin powered descent or abort. By pressing the proceed button, they'll trigger the firing of the descent rocket, which will gradually increase from ten to one hundred per cent thrust. During the descent from 50,000ft, Eagle's speed will slow from a peak of 17,000mph to just 2mph moments before landing.

The craft's landing radar will bounce signals off the Moon's surface to provide altitude and velocity data needed to help the crew land. When Eagle is 5ft from the surface, sensors on its legs will trigger a light in the cockpit to confirm touchdown.

There will be no radio contact when Eagle first fires its engine, as it will be on the far side of the Moon, and there will be no live TV pictures of the landing itself.

LUNAR LIGHT MYSTERY

A mysterious bright spot seen by the astronauts today has added fuel to a scientific debate about volcanic activity on the Moon.

Neil Armstrong and Buzz Aldrin both observed a light coming from the inner wall of the Aristarchus crater. Armstrong explained: "It seems to have a slight amount of fluorescence to it. The area in the crater is quite bright, with bright rays leading out of it on all sides."

The crater is located to the north-east of the face of the Moon and previous sightings from Earth of bright spots there have led some scientists to suggest this could indicate volcanic activity, although others believe the sporadic incidents are caused by thermoluminescence, a release of energy in the form of visible light.

CREW ENJOY A LIE-IN

The astronauts got a two-hour lie-in today, boosting the chances of an earlier Moonwalk.

NASA'S top doctor, encouraged by the amount of sleep the crew have been getting, has effectively given the green light for Neil Armstrong and Buzz Aldrin to skip a rest period after landing and bring the Moonwalk forward by several hours.

At 6.32am today, Aldrin checked in with Mission Control and was told a final mid-course correction, scheduled for two hours later, had been cancelled. A NASA spokesman said: "That was that. Apparently they turned over and went back to sleep!"

At a press conference tonight, Doctor Charles Berry, Medical Director, said: "We were quite concerned that if they didn't get adequate sleep on these first three nights, they'd face the lunar surface activity period when they were already fatigued. If there was a request on their part to alter the flight plan, we'd make a real-time medical decision based on whether we thought they were rested enough to undertake the EVA (extra-vehicular activity) at that time."

The astronauts are unlikely to be comfortable during the rest period in the lunar module because they'll be wearing bulky Moon suits and there are no seats in the cabin. One concern for Berry is the danger of Armstrong or Aldrin suffering from space sickness, which has been an issue for some astronauts. The pair will be the first men to stand upright and walk in gravitational conditions totally different to Earth and Berry said it would be "bad news" if one of the astronauts were to vomit while wearing a space helmet. They could even choke to death, he admitted.

Despite rigorous physical and psychological tests, Berry believes some types of motion sickness are only triggered in particular situations in space. He said: "This makes astronaut selection very hard. I don't think we have anybody who is prone, but there are gradations of sensitivity among the astronauts."

PAIR GO INTO ISOLATION

Two volunteers went into quarantine today ready for when the astronauts return to Earth.

Doctor William Carpentier, a Canadian-born physician, and Doctor John Hirasaki, a Texan scientist, are in lockdown in the Mobile Quarantine Facility on board the USS *Hornet*. They'll be quarantined with the astronauts until 11 August.

Carpentier and Hirasaki started their isolation early to reduce the risk of picking up any illness from members of the *Hornet*'s crew. Hirasaki confessed to feeling "a little edgy", while Carpentier predicted: "The quarantine promises to be very dull and I've no doubt we'll all be snapping at one another before we get out."

Carpentier has taken part in eight previous recovery exercises and said: "People always ask me what the astronauts say after splashdown.

The Mobile Quarantine Facility, which is on board the USS *Hornet*. A doctor and a scientist went into isolation inside the trailer today to prepare for the astronauts' return to Earth.

Actually, because of the noise of the sea and the helicopter I don't hear a damn word!"

His wife Wilma – mother to their two pre-school-age children – admitted the amount of time Carpentier had spent working on the space programme since 1965 had been difficult. She said: "I cry a lot. I can hardly wait until he gets home." Asked to describe her husband, she added: "He's intellectual and emotional. I wouldn't really know if he's romantic because I haven't seen him that often."

SOLAR DANGER RULED OUT

Scientists have given a perfect forecast for tomorrow's landing – lunar hailstorms, but no deadly solar flares.

Occasionally, balls of energy erupt from the Sun, releasing showers of X-rays and protons which would be fatal to humans. Stefan Pinter, a helio-geophysicist based in Czechoslovakia, had urged NASA to postpone the mission because he feared it could coincide with a bout of high solar activity.

But NASA was assured today that the astronauts won't be at risk. Seven stations around the world are constantly monitoring the Sun and can predict such an event forty-eight hours before it happens. Major Gerald McCright – who heads a solar observation team in Greece – confirmed they have seen nothing to suggest any threat to the mission.

The astronauts will, though, have to contend with continuous hailstorms on the surface. Doctor Elbert King, Curator of the Lunar Receiving Laboratory in Houston, explained that the hail is made up of micrometeorites; tiny pieces of rock and metal which burn up as shooting stars and hit the Moon at speeds of up to forty-five miles per second.

King said Armstrong and Aldrin's spacesuits are designed to withstand hits from micrometeorites and that the debris will explode on impact in flashes of light, which will look like fireflies to the astronauts. King said: "It's doubtful they'll even feel the hits."

In theory, a bigger micrometeorite could make a hole in a spacesuit, which would be catastrophic, but King insisted: "They could walk over hundreds of square feet of the Moon's surface in a day without running such a risk."

NEW SOVIET RIDDLE

Space experts were left puzzled today when the Soviet Union announced its Moon probe has switched into a higher orbit.

In a surprise announcement, *Tass* confirmed that, after twenty-six orbits at an altitude of between 34.5 and 126.5 miles, Luna 15 is now lapping the Moon at an altitude ranging from 59 to 136 miles.

Sir Bernard Lovell, Director of the Jodrell Bank Observatory in England, said a sixty-four-minute burst of data transmissions his team intercepted might even mean the probe has already divided into two sections, with one having landed. He said: "You can't absolutely exclude the possibility that it might have been transmitting to a capsule on the surface."

Later, *Tass* carried an interview with space engineer Anatoly Koritsky in which he referred to the probe being able to carry out research into subjects including "the chemical composition of lunar rocks". This reference could mean Luna 15 *will* attempt to bring back samples.

GLENN IS SO JEALOUS

The first American to orbit the Earth admitted today he is jealous of the Moon crew.

John Glenn became a national hero with his pioneering 1962 flight on Friendship 7, circling the planet three times at speeds of up to 17,500mph.

Glenn, who quit the space programme in 1964 to pursue business and political interests, followed today's events from the VIP viewing area at Mission Control and, as Apollo 11 prepared to go into lunar orbit, he said: "I'm green with envy. I'd give anything to be up there right now. Who wouldn't?"

The former Marine Corps pilot, who was 48 yesterday, said the strides made by the programme since his flight made him feel "as ancient as the Wright brothers". He added: "There's no comparison between my flight and this one. It's like comparing a horse and wagon with a Cadillac. If I were restored to flight status tomorrow, I'd need at least a year of studying to catch up."

Glenn said people shouldn't underestimate the courage of the astronauts or the dangers they face. He said: "They do have fears, of course. They're not, as some people apparently think, robots – something mechanical turned out by computers – or chimps doped up to be unafraid.

John Glenn (centre), the first American to orbit the Earth, has been following the progress of Apollo 11 at Mission Control. Here he is pictured with Bob Kline (left), chief of the Mission Operations Procurement Branch at the Manned Spacecraft Center, and Doctor Eberhard Rees, Deputy Director of the Marshall Space Flight Center.

"They've come through life along a different path than most people, usually as test pilots, and danger isn't new to them. The only thing abnormal about them is their curiosity. Their training and knowledge reduces their fears. There are, of course, moments of apprehension. Sitting up there on that huge rocket, you know what can happen if the thing blows."

HOUSTON PLAYS IT COOL

People in Houston are taking the mission in their stride as their local heroes prepare to make history.

Houston is home to the Manned Spacecraft Center (MSC) and the NASA astronaut team and support for the mission is clear for all to see in 'Space City'.

In Nassau Bay, where two of the crew live, the Resort Motor Inn has a sign declaring: 'Hail Columbia and Eagle, too … hiya Moon and see ya soon!' Across the road, the message at the King's Inn is: 'Let your flags fly for our heroes in the sky'.

Former astronaut John Glenn said: "The men are put on some kind of pedestal. Somehow, once we've flown, we aren't supposed to wash our car or take out the garbage. I've even had letters asking me what stocks to invest in. Suddenly we're supposed to be experts in everything."

However, there's a feeling among locals that Apollo 11 is just another mission. While half the USA's private employers are now expected to give staff a day off on Monday, most Houston stores will open and the city's trash collectors have voted to work for holiday pay.

Local architect Seth Morris said: "Every time one of those astronauts says 'Hello, Houston' we're all reflected in the glory. But I can't say we pay much attention unless there's a flight in progress."

Houston-based columnist Maxine Messenger recalled how astronaut Gene Cernan was the centre of attention at an event she attended in Las Vegas. She said: "Hundreds of movie stars were there and they all wanted to meet an astronaut. They fall all over them. It's like that everywhere I go. But to us, they're just plain folk."

The city is likely to hold a parade in honour of Apollo 11, but Lee Dubow, a Vice President at the city's Foley's Department Store, said: "We're not being blasé, but I think a lot of us can't see ourselves marching down the street for our neighbours and friends."

NASA's decision eight years ago to build the MSC twenty-five miles from downtown Houston has transformed the south-east of the city. It cost $2 billion, including land purchases, but William Lilly, Assistant Administrator for Administration, said: "It'll be in use long after the Apollo programme."

More than 50,000 people live in nine towns which make up the Clear Lake area and land prices along NASA Highway 1 have soared. Unlike in Cape Kennedy, confidence is high that the boom will continue. For every hundred jobs created by NASA, another sixty-five have sprung up in private commerce and industry and Ben Blackledge, Clear Lake Chamber of Commerce President, said: "There could be a levelling of the dynamic growth of the area, but the area was ready to boom anyway."

Jack Lister, Director of Personnel Services at the MSC, said: "There hasn't been much concern, because we're in the prime of our effort. The landing is our culmination and we're not doing too much thinking about afterwards. We're simply too busy."

HI-TECH POPE

The Pope is following the Moon mission in colour while the rest of Italy watches in black and white.

Italian TV has yet to upgrade its broadcasts to colour, but Vatican officials confirmed today that Pope Paul VI has arranged for a special circuit to be fed to his summer residence at Castel Gandolfo so he can watch in colour.

HAIRCUTS OUT OF THIS WORLD

Thr Moon men have been offered a free haircut by the 'official' outer space barber.

James Wallbridge, from Ellensburg, Washington, filed an application with Kittitas County Court in 1959 to operate a chain of salons on the Moon and Mars. The 61-year-old also laid claim to a franchise for 'an outer-planetary space-flight transportation and freighting company' to transport equipment and passengers.

He said: "Sure, it sounds fantastic, but it isn't. Wait and see. If I didn't think I'd be around to see it happen, I'd never have done it. The Moon will be a stopping-off place for space travel and barbers will be needed. Once the Moon starts being developed, the idea won't sound nearly as fantastic. Once the astronauts have been up there and bring back some of that Moon dust, there's nothing that's going to stop them from developing that place, not so long as there's a single taxpayer alive."

Wallbridge said he would be happy to give Neil Armstrong, Buzz Aldrin and Michael Collins a free haircut when they return to Earth and promised: "I would gladly share my Moon shops with them."

Asked about the possibility of a dispute with Chilean lawyer Jenaro Gajardo, who claims to be the legal owner of the Moon, he added: "I'd have to see his documents before I believe it. I can produce *my* proof."

SURPRISE TV STAR

Former astronaut Wally Schirra has denied he's guilty of double standards after becoming a star turn on TV.

Schirra, who made three spaceflights, is proving a hit as an expert analyst on CBS, which is a surprise to those who had him down as an opponent of TV's role in the space programme.

When he commanded Apollo 7, a tetchy Schirra postponed the first scheduled telecast, telling ground control: "We haven't eaten. I have a cold. I refuse to foul up our timelines."

It was widely reported that he was unenthusiastic about the introduction of TV cameras into the cockpit, but today he insisted: "I want to dispense with that image. You know, 'Here's the guy who hated TV now working for a TV network'. In contrast, I fought hard to have TV aboard. I said the world needs to see a mission as we do. Some people were worried about the Russians reading our instruments. I wasn't worried."

'DON'T FORGET SPUTNIK SHOCK'

Former Vice President Hubert Humphrey today urged opponents of future space exploration not to forget how America was shown up by Sputnik.

Humphrey, who lost last year's election to Richard Nixon, is a former Chairman of the National Aeronautics and Space Council. Recalling the impact of the Soviets' first satellite launch in 1957, he said: "A new star appeared, the red star of the Soviet Union, the red star of Sputnik. I remember the shock that spread across America when we realised the Russians were first in space.

"We realised that the competition from the Soviet Union was much tougher than we had anticipated, not only in ideology, but also in technology. Sputnik forced us to re-examine our educational system. It made us take a hard look at our industrial capacity and national security.

"If the Apollo 11 flight goes well, we will be, unquestionably, the world leader in space exploration and technology. The space programme has been hotly debated. I'm convinced it's a wise investment in the future. It's more than just an adventure."

He cited numerous benefits "behind the fanfare of the missions". These included advances in medicine, communications, weather forecasting, electronics and computers. He added: "The programme, more than any other single development, has upgraded American industry and management. The space programme is just another name for excellence in American technology. It has helped fuel our fantastic economic growth."

Humphrey, who is currently on a private visit to the USSR, said he would be stressing to Soviet officials the need for co-operation in future space ventures.

MARS PLANS 'PREMATURE'

The head of NASA today played down the chances of a manned mission to Mars in the near future.

As NASA Administrator, Thomas Paine is a key member of the Space Council tasked with making recommendations to Richard Nixon over future plans. In an interview on the eve of the lunar landing, he insisted: "We have no plans at the present time for a Mars expedition and no crash programme to get such plans together. We view the exploration of the solar system as being a job that will exercise men for many, many generations to come. A Mars expedition in the 1970s is probably premature. The real question is what the Moon programme should be beyond the first ten trips. It's a very difficult set of trade-offs and decisions."

Paine believes a Moon-orbit station, allowing astronauts to commute to and from different locations on the surface, would be more practical than a base on the Moon. But he feels an Earth-orbit station and a reusable shuttle should be the first priorities. He said: "One of the fundamental points of the second decade of space has got to be innovations that will drastically bring down the cost. A space shuttle vehicle, capable of going from the Earth into orbit and back again, is one of the keys to low costs. The key is the ability to fly this vehicle up and back, refuel it and use it over again. This is quite an engineering challenge to us. Our guess is that we're right on the threshold of the technology that will enable us to do this."

Paine says an Earth-orbit station could eventually accommodate up to 100 people. He added: "These wouldn't be astronauts. These would be some of the world's foremost scientists carrying out experiments in astronomy, high-energy physics, materials and programmes of biology and medicine, programmes looking back on Earth, studying the weather, the oceans, the atmosphere, resources, geology. They would be scientists and engineers from all over the world."

PREACHER'S SPACE MISSION

Evangelical Christian leader Billy Graham believes there could be life on Mars – and he'd like to travel there to spread the word of God.

The influential preacher and civil rights campaigner insists space exploration should strengthen people's religious faith. He said: "Human achievements of science don't diminish God. They glorify God through the accomplishments of men ordered by God. For almost 2,000 years, the laws of science have changed dozens of times. God's law never changes."

Minneapolis-based Graham, a close confidante of Richard Nixon, added: "I'm often asked if I'd like to go the Moon. I would not! I want to go where there's life. I want to go where I can meet some of God's creatures. Perhaps life exists on Mars. Yes, I'd like to go there. What would I find? I've no idea. The Bible doesn't say, but it's hard to believe that we earthlings are alone in this wonderful universe.

"Already we've received visits by creatures from space, including many angels and Jesus Christ. And then there was – and *is* – Lucifer, the Devil. Space scientists have gone to great lengths to prevent contamination of the Earth, but we're already contaminated. The germ Lucifer brought is called sin."

ASTRONAUTS HAIL ATLANTIC ROWER

The Apollo 11 crew have congratulated Englishman John Fairfax after he today became the first person to row solo across the Atlantic.

Fairfax, 32, landed at Fort Lauderdale, Florida, 180 days after setting out from the Canary Islands on his 4,000-mile journey. Sunbathers applauded him as he came ashore with girlfriend Sylvia Marrett, 30, who had waded into the surf to embrace him.

Fairfax said: "I wanted to do it and I did it. I feel bloody happy about it! I think it's a bit symbolic that I arrived in America, which is a son of England, you know … rather a large son! And I did it at the same time you Americans were reaching the Moon."

Among the messages of congratulation he received was a letter from the astronauts saying: "We who sail the new ocean of space pay our respects to the man who, single-handedly, has conquered the still-formidable ocean of water. As fellow explorers, we salute you."

Fairfax said he had expected to make the crossing in his self-righting boat *Britannia* in ninety days, but it ended up taking double that time. After rowing ashore while singing *Rule Britannia*, he admitted: "I'm damned glad it's over. I'm fed up with rowing!"

His girlfriend said: "He looks marvellous. A little thinner, I think." Fairfax said: "I didn't think much about women. If I had, I'd have gone crazy!"

CHAPTER 20

SUNDAY, 20 JULY
Launch Day Plus Four

"For one priceless moment in the whole history of man, all the people on this Earth are truly one."

US President Richard Nixon

"I was leaning against a wall, or maybe it was leaning against me! I cried – it seemed like forever – tears of relief. I'm still not able to believe it."

Joan Aldrin

A bootprint in the lunar soil photographed during today's historic Moonwalk.

FOOTPRINTS IN HISTORY

Man is on the Moon.

At twenty seconds past 10.56pm, Neil Armstrong planted his size nine-and-a-half boots on the lunar surface and declared: "That's one small step for man, one giant leap for mankind."

With footprints that will be preserved on the surface of the airless Moon for 500,000 years, he took his place in history alongside the world's most iconic explorers. Around nineteen minutes later, Buzz Aldrin joined him and described the alien landscape as "magnificent desolation".

The Moonwalk came after Armstrong pulled off a heart-stopping landing on the Sea Of Tranquility. With no TV pictures available during the landing, the world – including Mission Control – had to follow it in audio only. NASA flight technicians collectively held their breath as Armstrong took manual control of the lunar module and skimmed across the surface in a ninety-second search for a safe spot to land. Finally, Eagle touched down and there were cheers in Mission Control as Armstrong reported: "Houston. Tranquility Base here. The Eagle has landed." NASA confirmed the official time of landing as 4.17pm and 40 seconds.

Hundreds of millions of TV viewers and radio listeners tuned in to the drama. The Moonwalk came six hours, thirty-nine minutes after touchdown and was brought forward by more than three hours, making it a viewer-friendly time across the USA. In China, a news blackout of the mission meant around a quarter of the world's population were unaware of events.

The astronauts spent more than two hours walking on the Moon and it was transmitted live in grainy black and white images once Armstrong deployed a TV camera on his way down the lunar module ladder. Viewers saw the pair collecting samples, setting up scientific experiments and planting the Stars & Stripes. The astronauts paused for a brief chat with Richard Nixon.

When they climbed back inside Eagle and Aldrin closed the hatch, there was a round of applause at Mission Control and, once the astronauts settled down for an uncomfortable night in the cramped cockpit, NASA officials reflected on events. Eugene Kranz, the Flight Director who oversaw the landing, said: "I was speechless the last ten seconds before the landing and speechless a minute afterward." Doctor Kurt Debus, Kennedy Space Center Director, said: "This has been the most exciting night of my life."

Tomorrow, after a twelve-hour rest and preparation period, Armstrong and Aldrin face more critical moments when they attempt to take off and rendezvous with the command module. Eagle's descent stage will act as a launchpad and the astronauts know they are relying on the ascent engine to fire perfectly or condemn them to being stranded on the Moon.

PILOT'S SKILL SAVED LANDING

NASA's top boss admitted tonight that Neil Armstrong's skills as a pilot saved the landing from disaster.

Armstrong took manual control for the final stages of the descent and, realising the original target zone was too dangerous, had to skim across the surface until he identified a safe spot. The former test pilot – aware that if he used too much fuel he and Buzz Aldrin wouldn't have enough to take off again – eventually brought Eagle down around four miles from the proposed landing site. Armstrong told Mission Control: "That may have seemed like a very long final phase, but the auto-targeting was taking us right into a football field-sized crater with a large number of big boulders."

Eagle came to rest at an angle of four and a half degrees – it could have been at up to thirty degrees without any danger of toppling over – and is in an ideal position for take-off.

At a press conference tonight, Doctor Thomas Paine, NASA Administrator, conceded it was "extremely lucky" that Armstrong found a safe spot. He said: "As they looked down at the terrain they were originally headed down into – which we had thought was a safe place to land – it certainly indicated that we might indeed have had considerable difficulty if it had been a completely programmed landing. We knew the possibility of considerable roughness. That's why, in training, we emphasised the astronauts flying the last few feet and selecting a safe landing position."

At one point during descent, communications were blacked out. Asked if there was any anxiety in Mission Control, Paine said: "I wouldn't say we weren't anxious. In fact, I don't think we were anxious about that alone. We were *very* anxious."

The astronauts sounded calm, but Paine said medical data suggested "emotion was running very high, as it was in all of us". Armstrong's heart rate had jumped from 77bpm to 156 by the time he landed.

The day of destiny began when the three astronauts were woken at 7am by Capcom Ron Evans after managing about six hours sleep in the command module. Columbia, travelling at 3,660mph, had just emerged from the far side of the Moon on its tenth orbit.

Armstrong and Aldrin put on their Moon suits and crawled into the lunar module to prepare for separation. After Columbia disappeared behind the Moon on its twelfth orbit, Michael Collins released latches connecting the two craft and, when radio contact was re-established at 1.50pm, there was relief in Houston as Armstrong reported: "Eagle has wings."

At this point, Eagle and Columbia were flying only a few feet apart. But, twenty-two minutes later, Collins fired his manoeuvring rockets to put two miles between them. As Eagle receded into the distance – with Armstrong and Aldrin riding feet first and facing downward – Collins told them: "It looks like you've got a fine-looking flying machine there, Eagle, despite the fact you're upside down."

At 3.08pm, Eagle's guidance system automatically fired the descent rocket for 29.8 seconds to slow the craft so that it fell out of its sixty-nine-mile-high orbit and began a long, curving trajectory down to within 50,000ft of the surface.

Once again, the two craft were out of radio contact with Houston and there was near-silence at Mission Control as technicians, NASA top brass and current and former astronauts waited anxiously to hear if Eagle's engine had fired at the right moment and for the correct amount of time.

At 3.47pm, Collins assured Capcom Charlie Duke: "Listen, babe, things are going just swimmingly … beautiful!" After two minutes of receiving static noise from Eagle's radio, controllers heard Armstrong say: "The burn was on time." Eugene Kranz, Flight Director, told Mission Control staff: "We're off to a good start. Play it cool."

By 4.05pm, fifty-seven minutes after Eagle dropped out of its original orbit and with only 50,000ft to go, the powered descent phase began, with a gradually-increasing thirteen-minute burn of Eagle's 9,850lb-thurst descent rocket acting as a brake.

Five minutes later, 36,000ft from the surface, the astronauts passed up a last chance to abort and rendezvous with Columbia using only their descent engine. Below this point, they would have had to use up their descent fuel, jettison the descent stage and use the ascent engine for rendezvous.

Armstrong and Aldrin were unable to see where they were heading but, at an altitude of 7,200ft, Eagle's control jets fired to begin tipping the module into an upright position. With three and a half minutes to go and Eagle descending at 20ft per second, the astronauts could see the Moon's surface. When their altitude was 3,000ft, Duke told them: "You're go for landing!"

At 300ft, Armstrong took manual control and, realising they were heading for a crater, knew he had little time to identify a safe spot before he'd have to abort and blast back up using the ascent engine.

Eagle was dropping at two-and-a-half feet per second and, with Aldrin giving him constant updates on their altitude and other data, Armstrong fired thrusters to skim over the crater and onwards. Eventually, spying a clear, flat area, he brought Eagle down with forty-five seconds-worth of fuel left.

As the craft stirred a cloud of dust, a blue light in the cockpit flashed to indicate the probes on the lander's legs had touched the surface. Aldrin reported "Contact light" and then confirmed: "Okay, engine stop."

A tense sixteen seconds ticked by before Armstrong said: "Houston. Tranquility Base here. The Eagle has landed." As applause broke out in Mission Control, Duke responded: "Roger, Tranquility. We copy you on the ground. You got a bunch of guys about to turn blue. We're breathing again. Thanks a lot."

While the astronauts ran through system checks and the procedures for an emergency take-off, Kranz asked each of his frontline controllers for a "stay or no-stay" decision. Once the astronauts had been told it was safe to remain, Armstrong gave the world a first description of the scene outside. He reported seeing "a relatively-level plain" featuring "a fairly large number" of large craters and "literally thousands" of smaller ones. He said the surface was "pretty much without colour".

Aldrin said: "It looks like a collection of just about every variety of rock, every variety of shape, angularity, granularity. The colour varies pretty much on how you're looking. It looks as though some of the rocks and boulders are going to have some interesting colours."

Then, looking up through the overhead hatch, Armstrong said: "I'm looking at the Earth. It's big and bright and beautiful."

TOO EXCITED TO SLEEP

US TV networks got their way when the Moonwalk started early, but NASA insists it only happened because Neil Armstrong and Buzz Aldrin were too excited to sleep.

When the decision came tonight to scrap a planned rest period and bring forward the walk, Capcom Charlie Duke joked: "You guys are getting primetime TV!" However, Doctor Charles Berry, Medical Director, insisted the change was entirely down to the astronauts. He said: "It would've been virtually impossible for them to go to sleep when they were excited."

For viewers in the east of the USA, Armstrong's first step came at 10.56pm, while it was 7.56pm on the west coast. In Europe, people expecting to watch at breakfast time had to stay up into the early hours. In the UK, for example, it was 3.56am.

When, about two hours after landing, the astronauts radioed asking Mission Control for permission to bring the walk forward, they were immediately given the go-ahead. Preparations began around three hours

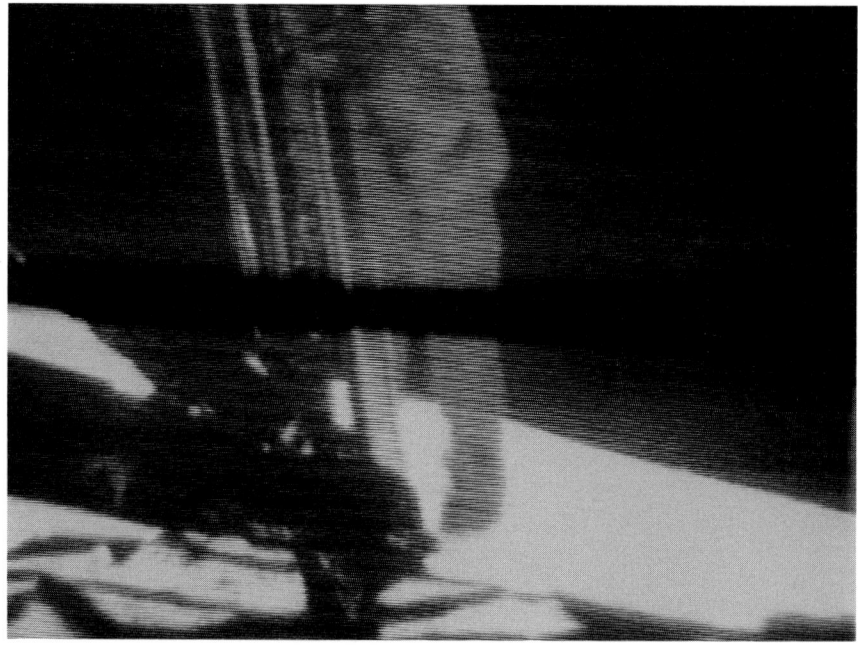

The grainy image TV viewers saw tonight as Neil Armstrong stepped off the ladder of the lunar module.

later when the astronauts – who by now had eaten man's first meal on the Moon – started cabin depressurisation. After a tense wait of almost an hour, Armstrong opened the hatch. As he got ready to wriggle through the opening in his bulky Moon suit, he said: "Now comes the gymnastics!" With some guidance from Aldrin, he made it. "Okay, Houston. I'm on the porch."

Mission Control and TV viewers had been following events in audio only, but when Armstrong reached the second of the nine rungs on the lunar module ladder, he deployed a TV camera that had been safely housed inside a folding compartment.

There were cheers at Mission Control as the first images flickered on to their monitors. Bruce McCandless, who was now on Capcom duties, told him: "Okay, Neil, we can see you coming down the ladder."

Moving cautiously, it took nearly a minute before he got to the bottom of the ladder and reported he was ready to step off. Then, with a little hop, he was on the surface and declaring: "That's one small step for man, one giant leap for mankind."

"The surface is fine and powdery. I can pick it up loosely with my toe. It does adhere in fine layers like powdered charcoal to the sole and sides of my boots. I only go in a small fraction of an inch, but I can see the footprints of my boots and the treads in the fine, sandy particles.

"There seems to be no difficulty in moving around. It's absolutely no trouble to walk around. The descent engine didn't leave a crater of any size. We're essentially on a very level place here."

As Armstrong got swept up in the excitement, he forgot to scoop up a contingency sample he was supposed to grab in case he had to abort the Moonwalk. Instead, he asked Aldrin to lower him down a still camera. Armstrong busied himself taking photographs and it was eight minutes before he collected the sample, having been prompted several times by McCandless.

Armstrong found it harder than he expected to dig below the topsoil. "This is very interesting. It's a very soft surface, but where I plug with the sample collector, I run into a very hard surface." Describing the landscape, he explained: "It has a stark beauty all its own. It's like much of the high desert of the United States. It's different, but it's very pretty out here."

When Aldrin's turn came, he clambered out of the hatch and said: "Now, I want to partially close the hatch, making sure not to lock it on my way out. That's our home and we want to take good care of it!"

Aldrin jumped from the fourth rung of the ladder and, as he bounced along in the one-sixth gravity, exclaimed: "Isn't that something!"

Once they had set up a TV camera on the surface to show the lunar module and surrounding area, viewers were able to see the astronauts moving around in bounding steps. Aldrin explained: "You do have to be rather careful to keep track of where your centre of mass is. Kangaroo-hop does work, but it seems as though your forward mobility isn't quite as good as it is in the more conventional one foot after another. Got to be careful that you're leaning in the direction you want to go, otherwise you appear slightly inebriated."

Armstrong unveiled a plaque attached to Eagle's descent stage and described it to the world. "There's the two hemispheres. Underneath, it says 'Here, men from the planet Earth first set foot upon the Moon. July 1969 AD. We came in peace for all mankind'. It has the crew members' signatures and the signature of the President of the United States."

When planting the Stars & Stripes, the astronauts needed five attempts to make it stand upright. Once they did, the pair stood and saluted the flag.

Armstrong and Aldrin scurried around, setting up three scientific experiments and collecting samples. Two airtight boxes were filled with rocks and soil.

Mission Control granted an extra fifteen minutes on the surface to complete their tasks and one of the final jobs carried out by Aldrin was to hammer sample tubes into the ground to capture material from beneath the surface. Struggling to penetrate the surface, he said: "I hope you're watching how hard I have to hit this, Houston."

When time was up, Aldrin clambered back into Eagle, calling out to Armstrong: "Adios, amigo!" The Commander followed about twelve minutes later and the hatch was closed at 1.12am, two hours and thirty-two minutes after it was first opened. Astronauts' chief Deke Slayton came on the Capcom to tell them: "That was a real great day, guys! I really enjoyed it." Armstrong replied: "You couldn't have enjoyed it as much as we did!"

Assessing events, Doctor Thomas Paine, NASA Administrator, said: "The voices we hear coming back from these brave men on the Moon are hard to believe, but it has raised the spirits of men around the world. We have demonstrated a very crude and preliminary form of travel between the Earth and other bodies. We've entered a new era. The significance of the trip that has just started is indeed that mankind is going to establish places of abode outside of his home planet."

NIXON PHONES THE MOON

Richard Nixon chatted with the astronauts on the Moon tonight and told them: "This certainly has to be the most historic telephone call ever made."

The President's two-minute conversation with Neil Armstrong and Buzz Aldrin was screened live across the world. The signal was transmitted from the White House switchboard to Goddard Space Flight Center in Maryland, on by radio link to Houston and then through the Goldstone antenna in California to the Moon. Viewers saw Nixon in the top left corner of their screens and the astronauts standing either side of the US flag.

Speaking from the Oval Office, Nixon said: "For every American, this has to be the proudest day of our lives. For people all over the world, I'm sure they, too, join with Americans in recognising what an immense feat this is. The heavens have become a part of man's world. As you talk to us from the Sea of Tranquility, it requires us to redouble our efforts to bring peace and tranquility to Earth. For one priceless moment in the whole history of man, all the people on this Earth are truly one; one in their pride in what you've done and one in our prayers that you'll return safely to Earth."

Armstrong replied: "It's a great honour and privilege for us to be here representing not only the United States, but men of peace of all nations, men with an interest and a curiosity and men with a vision for the future."

At a late-night press briefing, Ronald Ziegler, White House Press Secretary, revealed Nixon had phoned his predecessor, Lyndon Johnson, and former First Lady Mamie Eisenhower straight after speaking to the astronauts. Dwight Eisenhower's widow reportedly told Nixon that "somebody up there is looking at them too" in a reference to her late husband. The President had followed the landing on TV alongside Frank Borman. Describing Nixon's reaction, Borman disclosed that the President had said: "It's an unbelievable thing. It's fantastic!"

In an ABC TV interview, meanwhile, Borman hit back at claims that Nixon's Moon call was a cheap publicity stunt. Borman insisted it wasn't Nixon's idea and said: "I'm surprised that anyone could object to his speaking to his explorers – to *our* explorers – on the Moon. We suggested that the President do just this. I guess you have to blame NASA, if you're going to blame anyone. I think it's not only appropriate but, as chief of state, it's mandatory that he did it."

LOST IN SPACE

Michael Collins endured a frustrating time tonight trying to find the 'lost' lunar module.

Collins, lapping the Moon alone in the command module, found himself engaged in a fruitless attempt to spot Eagle on the surface as Mission Control tried to establish exactly where it landed.

As a result of Neil Armstrong's manoeuvre to avoid a crater and a minor error in information fed to Eagle's on-board computer, NASA was unable to confirm the exact co-ordinates. Armstrong told Houston: "The guys that said we wouldn't be able to tell precisely where we are are the winners today. I haven't been able to pick out things on the horizon as a reference."

Collins was asked to look out for Eagle each time his two-hour-long orbit took him over the Sea Of Tranquility, but he said he'd had "no joy" in three separate attempts.

Earlier, Capcom Charlie Duke relayed the news of touchdown to Collins, who responded: "Yeah, I heard the whole thing. Fantastic!" When Duke told Armstrong and Buzz Aldrin there were "a lot of smiling

Technicians at Mission Control watch live TV pictures of today's Moonwalk on a giant screen.

faces in this room and all over the world", Collins quickly chipped in with: "And don't forget one in the command module!"

Collins was initially unable to speak to his crewmates. When the radio connection was re-established, he told them: "It sure sounded great from up here. You guys did a fantastic job."

The Moonwalk began just as Collins went behind the Moon and, when he re-emerged, he asked Mission Control: "How's it going?" Capcom Bruce McCandless explained that Armstrong and Aldrin were busy planting the Stars & Stripes flag. "I guess you're about the only person around that doesn't have TV coverage," said McCandless. Collins insisted: "That's all right. I don't mind a bit."

The lone astronaut was again out of contact when the Moonwalk finished. When McCandless informed him they were safely back inside the lunar module, Collins exclaimed: "Hallelujah!"

COMPUTER OVERLOAD DRAMA

An overload of the lunar module's on-board computer came within moments of wrecking today's landing.

Neil Armstrong and Buzz Aldrin were mystified when two programme alarm codes flashed up on their cockpit display during powered descent. Neither alarm had featured in the many scenarios thrown at them in training and only the in-depth knowledge of flight technicians in Houston saved them having to abort.

About seven minutes before touchdown, Armstrong reported: "Programme alarm. It's a 1202." As Eugene Kranz, Flight Director, waited for an answer from the floor at Mission Control, Armstrong demanded: "Give us a reading on the 1202 programme alarm." Just twenty-seven seconds after Armstrong first flagged up the problem, Capcom Charlie Duke assured him: "We're go on that alarm."

Four minutes later – and just moments after Kranz had given the final go-ahead for landing – another warning flashed up. Aldrin said: "Programme alarm. 1201." Almost immediately, Duke came back with: "Same type. We're go."

Tonight, Stephen Bales, Guidance Officer, explained that both alarms were the computer's way of saying it was struggling to carry out so many tasks simultaneously. Bales said: "The computer was working at a very high percentage and decided it just had too much work to do. As long as the alarm doesn't reoccur constantly, you're okay."

DOCTORS ARE DELIGHTED

NASA doctors tonight declared themselves delighted with how Neil Armstrong and Buzz Aldrin coped with the Moonwalk.

Doctor William Hawkins, Flight Surgeon, said the energy expended was comfortably within safety limits drawn up on the experience of training exercises.

Aldrin's peak heart rate was 125bpm. Armstrong's did rise to 160 at one point, but Doctor Charles Berry, Medical Director, said this only happened when he was loading two heavy boxes of lunar samples onto the lunar module. Berry said: "His heart was probably beating a lot slower than mine! We couldn't be happier with their physiological state right now."

LASER EXPERIMENT HICCUP

Scientists' first attempts to test a $2 million Moon reflector have flopped.

Minutes after the Moonwalk ended, Lick Observatory in California began firing laser beams at equipment set up on the surface. However, confusion over Eagle's exact landing spot meant scientists at Mount Hamilton were given the wrong co-ordinates.

By bouncing light off the lunar-ranging retro-reflector and measuring how long it takes to make the return journey, scientists hope to precisely measure the distance between the Earth and Moon. Rotational wobbles of the Earth and Moon are thought to be connected to seismic activity here and accurate monitoring of the distance between them could help predict earthquakes.

Joseph Miller, Assistant Astronomer at Lick, said lasers had been fired for sixty-five minutes until they were told to stop by the Federal Aviation Administration. Officials were concerned that, as the Moon got closer to the horizon, the powerful beam might endanger people living near the observatory.

Another experiment left on the Moon proved instantly effective. A miniature seismic station – left on the surface to monitor for moonquakes – immediately began to transmit data back to Earth. When Armstrong and Aldrin jettisoned unwanted equipment from the lunar module on to the surface after the Moonwalk, Capcom Bruce McCandless reported their 'trash' had registered on the monitor. Armstrong joked: "You can't get away with anything anymore, can you?"

Buzz Aldrin sets up a seismometer on the lunar surface during today's Moonwalk. NASA said the experiment had proved instantly effective.

A third scientific experiment carried out during the Moonwalk saw an aluminium foil sheet erected to capture tiny particles carried from the Sun to the Moon by solar winds. The foil is being brought back for analysis.

GEOLOGISTS THRILLED

Geologists believe the astronauts' near miss with a lunar crater could be a blessing in disguise.

Doctor Gerard Kuiper, a world-renowned astrologer and space scientist who advised NASA on potential landing sites, said rock samples collected from near a crater were more likely to answer scientific questions about the Moon because they may have been forced up from deep beneath the surface when the crater was formed.

Kuiper, Director of the University of Arizona Lunar and Planetary Laboratory, described it as "a scientifically perfect spot" to collect samples. He said: "Landing near a crater gives an unusual opportunity to collect rock samples which come from considerable depths in comparison to the surface rocks."

Describing some of the rocks as he collected them, Armstrong mentioned that some rocks looked like basalt, a fine-grained dark rock formed by lava flows. He also referred to crystals in some rocks, maybe indicating the material was once molten.

Doctor David Mackay, a NASA geologist, said: "I was impressed by the variety of rock fragments in the area. It appears we have a very impressive array of geological samples to be returned." However, he cautioned against rushing to conclusions based on the astronauts' descriptions. "You can't really say what it means without looking at a rock in detail."

FIRST WORDS A SURPRISE

Neil Armstrong's parents were as surprised as anyone at their son's first words on the Moon.

Armstrong had refused to be drawn over what he might say and Viola Armstrong said tonight: "We had no idea. But I think it was great. I could tell from his voice that he was pleased and tickled and thrilled. It looked like he was having fun."

The Armstrongs gave two press conferences outside their Ohio home, looking calm and relaxed when they spoke within an hour of the landing but then appearing particularly emotional when they appeared again partway through the Moonwalk.

Viola blew kisses to the media and well-wishers, while husband Stephen clasped his hands above his head like a victorious boxer. Talking about the landing, Viola said: "I was worried the Moon might be too soft and they might sink in too deep. But, no, they didn't. So it was wonderful."

Asked about her son's unique place in history, she insisted: "Neil doesn't like to be called a hero. He thinks other people should get more of the credit; people like Wernher Von Braun and others who made the trip possible. Neil's not going to change just because of this. Not if I know my son." Stephen said his son sounded like "just the same old Neil" as he spoke from the lunar surface and that it was "real nice" Richard Nixon had spoken to the astronauts.

Celebrations are on hold in Wapakoneta until splashdown. Viola added: "That's when we will know for sure that Neil's safe. I wish I could send a message up there. I'd tell him 'Speed it up a little, honey, and get back home'."

Reverend Herman Weber, of the St Paul United Church of Christ, watched the coverage with the family. Referring to Armstrong's first words on the surface, Weber said: "It was really thought-provoking, historic, even poetic."

The astronaut's sister, June Hoffman, 35, watched at her home in Menomonee Falls, Wisconsin. The mother-of-seven said: "I was pushing every button with Neil and Buzz during the landing and I felt every emotion imaginable. We were all on the edge of our chairs. We're all pretty excited right now and breathing a little easier."

NATION IS GRIPPED

Tens of millions across America were gripped by today's drama, from Disneyland to death row.

While many people stayed at home to follow TV and radio coverage, others shared the experience by gathering to watch on giant screens. In New York, an estimated 4,000 people braved heavy rain to watch in Central Park, with many observing a call to wear white in honour of the astronauts. Times Square, the Time-Life Building and the Rockerfeller Center also attracted large crowds.

Visitors to Disneyland in Anaheim, California, made a bee-line for a big screen in Tomorrowland. They included eighty members of the Soviet track and field team, who had competed in an athletics meeting at Los Angeles Coliseum yesterday.

So many people were glued to their TV sets in Memphis, Tennessee, that water board officials had to shut down the city's five pumping stations to head off a dangerous build-up of pressure caused by no one using any water for a seven-minute spell during the landing.

Normally busy bars and restaurants in major cities were unusually quiet. In San Francisco, Twilight Zone Tavern bartender Frank Silvestri, 45, said: "I guess everyone is home watching it on TV. The streets are kind of deserted." At Jorgensen's Holiday Inn in Helena, Montana, manager Tony Culum served free Champagne and got through twenty cases. He said: "We can't be on the Moon's surface to christen it there, so we're doing it right here."

Las Vegas celebrated in unique style, with strippers at the Silver Slipper Casino peeling off special spacesuit costumes. On Footprint Island, Florida, the whole population watched the landing; all five of them. Aria Gibbs, mother of the only family who live there, said: "It's one big thrill. We wouldn't have missed this for the world." In Seattle, Washington State, the Dial family's son was born just as the landing happened. He has been named Neil Armstrong Dial.

Among those who were working as events unfolded in space were emergency services personnel. After managing to catch the landing on TV, Richard Hunt, District Fire Chief for Toledo, Ohio, said: "This is just another work day for us. We've been watching in between our work. We sure are hoping the alarm bell doesn't go off at about the time they step on the Moon!"

Two students at Pacific Union College in Angwin, California, celebrated winning a national newspaper group's competition to guess the correct time Neil Armstrong set foot on the Moon. Paul Schoenwetter, from San Rafael, and Bruce Wesner, from Sonoma, were only five seconds out. The pair, who submitted thousands of entries for the prize of a holiday anywhere in the world, plan a skiing trip to Switzerland.

Many inmates at the nation's prisons also watched the coverage. At California's San Quentin, the 1,100 prisoners were given permission to watch TV after their usual 11pm curfew. Even Sirhan Bishara Sirhan – awaiting the death penalty for killing Senator Robert Kennedy last year – was allowed to watch in his cell on death row. At Nevada State Prison, Carson City, one inmate used the distraction of the landing to attempt an escape. But he was spotted climbing a fence and surrendered after guards fired two warning shots.

The Moon landing wasn't welcomed by all. Many of the 50,000 crowd at a soul music festival in Harlem, New York, booed when it was announced.

WIFE'S PERFECT NIGHT

Jan Armstrong watched her husband become the first man to walk on the Moon, then insisted it *wasn't* the most memorable day of her life.

The astronaut's wife was in buoyant mood as she spoke to 150 reporters outside her Houston home, saying: "It's fantastic and I'm just as excited as you all are. I was terribly excited." But asked if it was the greatest moment of her life, she replied: "No. When I was married, that was the greatest moment."

Jan, wearing a gold lunar module pin on her blouse, added: "The evening's been unbelievably perfect. It's an honour and privilege to share with my husband, the crew, the Manned Spacecraft Center, the American public and all of mankind this magnificent experience." When one reporter told her "the man in the Moon is your husband now", she joked: "And is he green?"

Jan followed the landing on TV and through her squawk box, with Apollo 8 astronaut Bill Anders pointing out locations mentioned during the descent on charts spread out in front of them. A NASA official assigned to the family said that, as Armstrong climbed down the lunar module ladder, Jan said: "I can't believe it's really happening."

Unlike the other two wives, Jan didn't attend church today, but Father Eugene Cargill, from the Shrine of the True Cross Catholic Church, visited the family this morning. He told reporters as he left: "Technically and scientifically, they've done all they can. When it gets down to the nitty-gritty, it's in the hands of God."

TEARS AND DISBELIEF

Joan Aldrin told tonight how she cried tears of relief when the lunar module touched down safely.

Apollo 9 astronaut Rusty Schweickart was at the Aldrins' home to talk Joan through events and Bill Der Bing, NASA liaison officer to the family, revealed that Joan shouted "Touchdown!" when Eagle landed. She then hugged her father, Michael Archer. Der Bing said: "There were whoops and hollers from everyone. There was a lot of noise! When it quietened down, someone said 'Resume breathing'."

In a press conference after the landing, Joan said: "I was leaning against a wall, or maybe it was leaning against me! I cried – it seemed

Buzz Aldrin leaves the lunar module to join Neil Armstrong on the surface. Aldrin's wife, Joan, admitted today she found it "hard to think it was real".

like forever – tears of relief. I'm still not able to believe it. It's more than the human mind can comprehend, especially mine!" Later, talking about the Moonwalk, she said: "It was hard to think it was real. I felt I was looking at another simulation."

Son Andy, 11, was asked if he'd like to go the Moon and said he hadn't decided yet. Amateur actress Joan feigned a look of horror and said: "Oh Lord! I didn't know you had it in mind. I couldn't stand this *twice*."

Before the landing, Joan attended church and went shopping at a supermarket. She appeared close to tears when, as she left church with her children, they had to push their way through a media crowd and run to their car.

BUZZ'S SECRET COMMUNION

Buzz Aldrin secretly took a portion of sacramental bread to the Moon so he could take communion there, it was revealed today.

The astronaut is a lay preacher at the Presbyterian Church of Webster and, at Sunday service, Joan Aldrin wept as Reverend Dean Woodruff held up the sacramental bread and told the congregation: "This loaf isn't complete. Two weeks ago, Buzz took a part of the loaf with him. Now we'll commune with him. He's one of us."

Three hours after today's landing, during a lull in the mission, Aldrin told TV viewers: "I'd like to take this opportunity to ask every person listening in, whoever and wherever they may be, to pause for a moment and contemplate the events of the past few hours and to give thanks in his or her own way."

COLLINS 'WITH CREWMATES IN SPIRIT'

Michael Collins' wife tonight dismissed suggestions he's frustrated at not walking on the Moon.

The Command Module Pilot is orbiting the Moon alone as Neil Armstrong and Buzz Aldrin write their names in history and some commentators suggested today that there was a hint of frustration in some of Collins' conversations with Mission Control.

But Pat Collins told reporters outside their home: "Don't you think he's probably with them in spirit? He doesn't sound a bit disappointed to me. I know how happy he's been to be part of the team."

She said the day's events had been "fantastically marvellous", adding: "I thought it was positively beautiful. I wasn't nervous. I was excited; *very* excited! As a matter of fact, those men up there are a lot calmer than I am."

JOY ON EVERY CONTINENT

The landing captured the imagination of an audience watching and listening in every continent.

In Europe, it was the early hours of the morning when the Moonwalk began, but among the many people who stayed up was Britain's Queen Elizabeth II. She watched at Windsor Castle with husband Prince Philip. They woke their four children, including five-year-old Prince Edward, to see it.

Thousands watched on a giant screen in London's Trafalgar Square, with dozens jumping into the fountain to celebrate. Police said one reveller had been taken to hospital with a broken leg.

Both the BBC and ITV stayed on air through the night and ITV's coverage included a phone-in during which one excited gardener asked whether samples of lunar soil could help him improve on the prize-winning 85lb pumpkin he recently grew.

Today's events were particularly special for the world's astronomers. Sir Bernard Lovell, Director of Jodrell Bank Observatory in England, said: "The moment of the touchdown was one of the moments of greatest drama in the history of man. I'm just speechless with amazement."

Lucien Coallier, an amateur astronomer from Laval, near Montreal, Canada, had his telescope trained on the Moon and reported seeing the shadow cast on the surface by the lunar module. Doctor Andrzej Marks, a leading Polish astronomer, enthused: "I'm so excited; my pulse is cosmic!"

In France, General Robert Aubiniere, the head of the nation's space research centre, said: "It's unbelievable, like a science fiction thriller. What a beauty!"

In joyous scenes repeated in cities around the world, thousands celebrated on the Champs-Elysees in Paris after listening to live commentary broadcast through loudspeakers. Rene Barjavel, French science fiction writer, said: "We burst with pride and tremble with emotion. We're like the child letting go of its mother's hand and crossing the distance between two chairs for the first time."

In West Germany, American soldier John McGee, 22, from New York, watched on a giant screen in Frankfurt. He said: "I felt like yelling out loud. I was really proud. I couldn't believe those pictures were coming from the Moon." Frankfurt restaurant owner Calka Ainwohner, 52, said: "I've been looking at it on TV for twenty hours straight, but I still don't trust my own eyes!" And in West Berlin, lawyer Jurgen Mohr claimed: "It's a great triumph over communism."

The US Embassy in Rio de Janerio set up large TVs for Brazilians to watch and, in Malaysia, 14,000 people crammed into the National Museum in Kuala Lumpur to watch on giant screens and enjoy a lunar exhibition. In Australia, it was already the afternoon of 21 July when the Moonwalk happened. In rain-soaked Melbourne, shoppers sheltered under umbrellas as they crowded outside TV stores.

Events were followed closely in some of the world's most remote locations. In arctic Norway, the midnight sun kept the sky bright through the night as native Laplanders sat around their campfires, listening on transistor radios and composing folk poems in tribute to the astronauts.

Coverage was also heard by some Sherpa families living in the shadow of the world's tallest mountain, Everest, while tribesmen in rural areas of Zambia listened on 5,000 transistor radios handed out as educational aids by the government.

Not everyone was happy. The indigenous Uriana people in Venezuela condemned the landing. Yolanda Gonzalez, of the Guajiro tribe in Maracaibo, predicted: "From now on, all will be evil." And Spanish artist Pablo Picasso said simply: "It means nothing to me. I have no opinion about it and I don't care."

LEADERS HAIL HISTORIC DAY

World leaders were quick to offer their congratulations to America and NASA tonight.

Japanese Prime Minister Eisaku Sato sent a telegram to Richard Nixon congratulating the USA on achieving "a symbol of twentieth century civilisation". Sato told reporters: "Splendid! They've done it! I never thought such an event would take place in my lifetime." Danish Prime Minister Hilmar Baunsgaard could barely contain his excitement, saying: "I've hardly recovered my breath. This is almost too much all in one night!"

British Prime Minister Harold Wilson expressed his "deep wish for a safe return at the end of what has been a most historic achievement in the history of man", and Queen Elizabeth II spoke of her admiration for the astronauts "whose exploits add a new dimension to man's knowledge of the universe".

New Zealand Prime Minister Keith Holyoake called the landing "a stupendous voyage of discovery", and Guyana Prime Minister Forbes Burnham said: "It's great to be alive on this day. Hats off to America!"

Most goodwill messages expressed hopes that the landing would be a catalyst for greater efforts to end war, suffering and poverty. South Korean President Chung Hee Park said: "I pray that this auspicious festive day of the century will mark a great milestone in realising an eternal ideal; peace and prosperity for mankind on Earth."

Pope Paul VI was among religious leaders to offer their congratulations, but he warned that there was an obsession with technology "perhaps even to the point of folly", and that "everything still depends on the heart of man".

BIG ROLE DOWN UNDER

Australian Prime Minister John Gorton admitted he was "blinded by science" after following today's drama at a NASA tracking station Down Under.

Two facilities in Australia – Honeysuckle Creek near Canberra and Parkes Observatory in New South Wales – are part of the worldwide communications network and were largely responsible for the TV transmissions of the Moonwalk.

Gorton was at Honeysuckle and, asked to describe events there, he said: "Wriggly green lines and various noises coming out of bits and pieces of equipment; I was blinded by the science. I salute all those who have contributed to this great, peaceful achievement."

Honeysuckle, which was opened two years ago to support the Apollo programme, has a 210ft-diameter radio telescope, while Parkes boasts an 85ft satellite dish. Both picked up signals from Eagle shortly after Neil Armstrong's first step on the Moon and, once NASA's antenna at Goldstone, California, went out of range, the world relied on Australia to provide live pictures. The signal received from Eagle's antenna was so strong that Armstrong and Buzz Aldrin didn't have to erect a 10ft 'lunar umbrella'.

Clifford Charlesworth, a Flight Director in Houston, praised the Australians' role, saying: "It's obvious the support they provided us was pretty outstanding. It'll be years before anyone can beat that TV spectacular."

MOSCOW APOLLO SNUB

The Moon landing barely got a mention today on Soviet TV and radio.

As Eagle touched down, Moscow Radio was broadcasting a review of the week's sports, while Moscow TV was screening a biographical film about an opera singer. Eventually, Moscow Radio did confirm the landing, but the brief report was six minutes into its news round-up.

On TV, it was the final item on the last news telecast of the day. Stories deemed more newsworthy included reports on the twentieth anniversary of the Polish People's Republic.

Voice of America made a live Russian-language broadcast of today's events, but the frequencies they used were blocked by the Soviets. Richard Cushing, Voice of America Director, said: "We were hopeful they'd permit broadcasts, but the jamming began shortly after our broadcasts began."

MIXED EMOTIONS IN VIETNAM

Hundreds of thousands of American troops in Vietnam listened to the landing in a special live radio broadcast.

More than 500,000 troops are serving in Vietnam under General Creighton Abrams and a spokesman for US Command said: "I don't think General Abrams took time out to listen, but everybody else out here did."

Servicemen reacted to the landing with a mixture of pride and pleas for Richard Nixon to end American involvement in the war. Private Tom Tipiew, from Green Bay, Wisconsin, said: "It's a great accomplishment, like going from the old world to the new world." Specialist Steve Hall, from Dayton, Ohio, added: "It's an outstanding feat that all Americans can be proud of. We've all been praying for the astronauts."

However, Private Thomas Hruby, from Maple Heights, Ohio, summed up the feelings of many when he said: "It's good to venture out into space like this, but I think right now we're spending entirely too much money on such a job. Our country should be more concerned about the guys over here; getting us home. We have more problems in our own country, too. We should clear everything up in our own homes, our own towns. Then we can spend money on things like this." Specialist Charles Comstock, from Oklahoma City, said: "It's good we got on the Moon, but I'd like to see us get out of Vietnam."

NOTHING TO SEE HERE

Tourists who headed to Neil Armstrong's home town to join the party were left disappointed – when they found there *wasn't* a party.

Hundreds of visitors descended on Wapakoneta, Ohio, but no celebrations will take place there until after splashdown.

Ted Scnell, a local tavern owner, said lots of people from out of town had come in asking where the celebrations were happening. Schnell, who went to school with Armstrong, said: "One even asked when the parade was going to start."

VISITORS FLOCK TO MISSION HUB

Thousands of people flocked to the home of Mission Control today to be close to the action as history was made.

Many locals chose to make their first visit to the Manned Spacecraft Center in Houston and a NASA spokesman said: "We estimate we have had about 20,000-plus visitors today."

George Landry has lived in Houston for twenty-two years but had never been to the centre. He said: "I should've come long before now. The Moon shot is a wonderful achievement and I think we should continue with the space programme. It's a wonderful thing."

The centre was the scene of an anti-poverty protest today by the National Welfare Rights Organisation (NWRO), with about seventy-five protestors waving placards with slogans including 'NASA shoots for the Moon – we want to shoot for cheap food', and 'Remember the hungry in America too'.

NWRO leader Hulbert James called for a halt to further launches until the government started "an unconditional war to eliminate poverty by 1976, the 200th anniversary of the nation".

WHITE HOUSE PRAYERS

More than 300 American government and diplomatic corps officials attended a church service at the White House this morning to pray for the success of the landing.

Richard Nixon and his family headed a congregation of 346 at what was the biggest religious service held there since the new President took office in January.

Mark Hatfield, Republican Senator for Oregon, read one of the prayers. The outspoken Vietnam War critic told the congregation: "Excite our imagination to transfer this genius of co-operation and spirit of teamwork to our many other needs, lest our success on the Moon mock our failures on the Earth. Even as our astronauts go to the Moon in the name of peace, our world aches from the pain of wars."

Frank Borman read the opening ten verses of Genesis, just as he and his Apollo 8 crewmates did in an emotional live broadcast from the Moon last Christmas. He admitted he felt more nervous reading it today than he had when he was in space.

After the service, there was an embarrassing moment for the President as he explained how valuable Borman's input was in helping him follow the mission. Nixon said: "It's like talking to Columbus after he'd been to the New World and asking 'Is Magellan going to make it?' If Columbus says Magellan is going to make it, it's pretty sure he will." After a pause, Nixon added: "I'm not sure whether Magellan came after Columbus?"

Borman assured Nixon that he had. Portuguese explorer Ferdinand Magellan's circumnavigation of the Earth began in 1519, fifteen years after Italian-born Christopher Columbus completed the last of his Spanish-sponsored New World voyages.

The US flag flies on the Moon today as Neil Armstrong works beside the lunar module.

SCOUTS SALUTE ARMSTRONG

A scout jamboree that was name-checked from space by Neil Armstrong greeted news of today's landing with huge cheers.

During a telecast on Friday, former Scout Armstrong sent his best wishes to the 30,000 camping at Farragut State Park, Idaho.

Today, there was unbridled joy among the Scouts as they listened on transistor radios. Rick Slunaker, from Bothell, Washington, said: "Neil Armstrong makes me feel 10ft tall."

And Roger Armstrong, from Ojai, California, predicted: "It's opening the door to man getting a new start elsewhere and a new chance. Man blew it here on Earth with smog and everything. Now he may be able to try again and maybe learn from his mistakes."

PROBE MAY LAND SOON

Soviet Moon probe Luna 15 tonight appeared to be preparing to attempt a landing after swooping into a lower orbit.

Tass announced the latest change in orbit, which took the unmanned probe to within ten miles of the surface, just minutes before the Apollo 11 command and lunar modules separated ahead of the lunar landing.

Sir Bernard Lovell, Director of the Jodrell Bank Observatory in England, said: "I don't think the Russians can leave it in that orbit for very long because it may collide with a lunar mountain. It could still be that it's going to land."

NO MARS BET FOR £10k WINNER

Lucky punter David Threlfall tonight picked up a £10,000 Moon bet cheque then ruled out putting money on a Mars landing.

Englishman Threlfall wagered £10 in 1964 at odds of 1,000-1 on man making it onto the lunar surface before January 1971. Minutes after touchdown he received his winnings from bookmaker William Hill live on TV.

Personnel officer Threlfall, 25, from Preston, Lancashire, said: "I've no idea why I did it. It just came to me one day. I think I've had my share now. I don't think I'll be putting any money on people getting to Mars."

ALL QUIET IN 'ROCKET CITY'

America's 'Rocket City' is putting celebrations on hold until the astronauts are safely home.

The streets of Huntsville, Alabama – home of the Marshall Space Flight Center – were quiet today as people stayed indoors to watch the mission.

Mayor Joe Davis has confirmed tomorrow will be a day's holiday for public workers in the city, but major celebrations will only be held after splashdown. Davis said: "It's the greatest thing that's happened in this century. I have a renewed pride that Huntsville has played such a major part. I'm very happy they've made a safe landing and completed the first half of their mission. Now I just pray they can make it safely on the second half, the return to Earth."

VON BRAUN DODGED BOMBS

A wartime politician tonight sent a telegram to Doctor Wernher Von Braun telling him he was glad he wasn't killed in bombing raids he ordered.

Von Braun was the mastermind behind the Nazis' rocket programme, with weapons including the V2 rocket produced at Peenemunde on Germany's Baltic coast.

Duncan Sandys, chairman of a British War Cabinet committee responsible for defending the country against rocket attacks, sanctioned air strikes on Peenemunde. One raid in August 1943 killed a number of rocket engineers, but Von Braun wasn't at the plant.

In his telegram, Sandys, 61, told Von Braun: "Warmest congratulations on your great contribution to this historic achievement. I'm thankful that your illustrious career wasn't cut short 26 years ago."

'BANISH NIXON INTO SPACE'

A militant African-American leader today slammed the Moon shot and said Richard Nixon should be sent on a one-way trip into space.

Eldridge Cleaver, Black Panther Information Minister, said: "I don't see what benefit mankind will have from two astronauts landing on the Moon while people are being murdered in Vietnam and suffering from hunger, even in the United States."

In a bitter attack on the US political system and the state of the nation, Cleaver singled out Nixon, long-time FBI Director J. Edgar Hoover and powerful Arkansas Senator John McClellan.

Cleaver said: "We have a war going on in the United States and the casualty rate's rising. Nixon's the number one pig, the spokesman for the power structure and, as such, he oinks for all his barnyard fellows. I'd like to see an astronaut team composed of Nixon, Hoover and McClellan going to Mars with a shortage of fuel or a weak booster rocket; anything to guarantee they wouldn't come back."

Cleaver is currently in Algiers, having fled the USA in November. He was on parole after serving eight years of a thirteen-year prison term for assault with intent to kill and his parole was revoked after a shootout between Black Panther activists and police in Oakland, California, in which Cleaver was wounded and one of his group was killed.

BASEBALL TRIBUTE

Four Major League Baseball matches were halted today in honour of the landing.

The New York Yankees' home match against the Washington Senators was tied at the top of the eighth inning when the umpires brought play to a stop. Yankee Stadium announcer Bob Shepard told the 32,933-plus crowd: "Ladies and gentlemen, your attention please. You'll be happy to know that Apollo 11 has landed safely on the Moon."

A huge cheer greeted the announcement, with Shepard's words "on the Moon" barely audible amid the din. A message also flashed up on the scoreboard saying: "They're on the Moon!".

In Philadelphia, play was interrupted for about five minutes, with the players from the Phillies and Chicago Cubs lining up in the centre of the field to join the crowd in silent prayer. There were similar scenes at Montreal's Jarry Park, where the Expos were playing the New York Mets, and at Atlanta Stadium during the Braves' game against the San Diego Padres.

APOLLO 1 WIDOW'S DELIGHT

Apollo 1 astronaut Ed White's widow said tonight she had never lost faith in the space programme after her husband's death.

Pat White was at neighbour Jan Armstrong's house to celebrate the landing and she told reporters: "I couldn't be happier. It's the culmination of a dream. Ed would have been thrilled to see this day come. He really believed in it. Neil and Ed came into the astronaut programme together. They trained together. They were friends and neighbours."

Asked if she'd ever lost faith in the programme after her husband and two fellow astronauts died in 1967, she insisted: "No. I kept thinking Ed wouldn't have wanted us to. He never did. I kept thinking about that and fighting off any bitterness. I stayed in Houston because we felt a part of the community. We've felt very much a part of the programme. This is a special mission for me."

'DEMOCRACY WON SPACE RACE'

Lyndon Johnson today claimed the successful Moon shot proves that democracy works better than communism.

The former President believes the key decision which paved the way to the lunar landing was to put the programme under civilian, rather than military, control. In an interview broadcast by CBS tonight, he told Walter Cronkite that John F. Kennedy was on the verge of appointing General James Gavin as NASA Administrator when the space agency was first established.

Johnson recalled: "I said 'I think the worst thing we could do would be to put a man with stars on his shoulder and a general's uniform in charge because it would frighten other nations and be a great disservice to our programme'. I can't think of anything that would have been more disastrous; if other nations had thought we were occupying space to try and control the world and place armaments in space and threaten the existence of the world."

Asked why the USA had beaten the Soviets to the Moon, he said: "My best judgment is that they don't have the system of government we have. The great strength of our programme is its openness. We have a better system."

Space enthusiast Johnson believes investment in the programme can run in tandem with spending on major issues at home. "I think we have enough money in this country to do all the things we need to do, and that includes space and that includes education, health and these other things. We have a lot of problems that haven't been solved. Just as we didn't face up to space in the 1950s, we didn't face up to a lot of other things. We have to find some way all of us can live together without fighting and without killing and destroying."

RADIO LISTENER DIES

A TURKISH man collapsed and died after becoming over-excited as he listened to radio commentary on the landing.

Local officials in the province of Tokat said Mustafa Alfin, 62, was vehemently opposed to the mission and became agitated as Eagle touched down. Alfin reportedly shouted "Oh Lord, this will be the end of the world!" before suffering a massive heart attack.

CHAPTER 21

MONDAY, 21 JULY
Launch Day Plus Five

"I was proud to be watching. I was proud of them and I was proud to be an American."

Florida Mail Clerk Russell Caldwell

"When they got on to the surface I said to my wife 'I wonder who's holding the cue cards?'"

Pennsylvanian Ray Reppert

The lunar module ascent stage approaches the command module today ahead of rendezvous. The photograph, taken by Michael Collins from Columbia, shows the lunar surface, with the Earth rising in the distance.

HOMEWARD BOUND

Apollo 11 was homeward bound tonight after Neil Armstrong and Buzz Aldrin survived another do-or-die moment – the first-ever blast-off from the Moon.

People around the world breathed a collective sigh of relief as the lunar module ascent stage soared back into orbit and successfully docked with the command module.

It was the first time NASA had attempted to launch a craft without the security of a concrete base or a metal gantry. Had the engine failed or had Eagle toppled over, Armstrong and Aldrin would have been doomed to die. But the lift-off went like clockwork and tonight Apollo 11 fired its main engine to begin the sixty-hour journey home.

At a late-night news conference in Houston, Lieutenant General Sam Phillips, Apollo Programme Director, said: "The men and equipment that are Apollo 11 have performed to perfection. Perfection isn't too strong a word."

At 1.55pm, using the descent stage as a launchpad, Armstrong and Aldrin fired Eagle's single 3,500lb-thrust ascent engine to leave the Moon. The feed from a camera set up on the surface had been turned off to allow the astronauts to switch to a different communications frequency for better voice and data transmissions, so Mission Control and people around the world followed the launch in audio.

Capcom Ron Evans told the astronauts: "You're cleared for take-off." Armstrong replied: "Understand. We're number one on the runway." Sixteen minutes later, Aldrin counted down the final seconds until he pressed the 'proceed' button to tell Eagle's guidance system to fire the engine. "Nine, eight, seven, six, five … abort stage … engine arm ascent … proceed." As the craft lifted off, Aldrin reported: "Beautiful … 26, 36ft per second up. Pitchover … very smooth … very quiet ride."

Evans repeatedly reassured the astronauts that data received at Mission Control indicated all was well. "One minute and you're looking good … You're go at three minutes. Everything's looking good … Four minutes. You're going right down the track. Everything's great."

In a reference to the main road serving the US east coast, Armstrong responded: "We're going right down US-1. Looking good here. It's a pretty spectacular ride."

The Moon's relatively weak gravity meant take-off was possible with a small rocket and Eagle blasted off at one mile per minute before quickly picking up speed. After seven minutes, eighteen seconds, the engine shut down and Armstrong confirmed: "The Eagle is back in orbit." Evans replied: "We copy. The whole world's proud of you."

Eagle coasted from east to west across the face of the Moon, slowly climbing to its target altitude of around fifty-two miles. Four separate manoeuvres using thruster jets raised and circularised Eagle's orbit and then slowed it down as it closed in on Columbia. By now, the lunar module's weight – 32,400lb at the start of the mission – was only 5,800lb, with most of its fuel used up and the descent stage left behind on the Moon.

The build-up to docking took place over the course of two orbits, including two tense periods without radio communications while the craft was behind the Moon. A commentary from John McLeaish, Chief of Public Information at Houston, kept TV viewers and radio listeners updated on the ever-decreasing distance between the two modules.

When Columbia and Eagle emerged from behind the Moon on the first occasion, Michael Collins told Mission Control: "Everything's going beautifully." McLeaish informed the worldwide audience: "The displays here show the range from Eagle to Columbia is a tad over 100 miles and about 99ft per second closure rate."

Flying over the Sea of Tranquility, Collins joked: "Eagle, Columbia. Passing over the landing site. It sure is great to look down there and not see you!"

Two hours later, with the craft set to emerge from behind the Moon for a second time, McLeaish reported: "Less than a minute now away from acquisition of the spacecraft Columbia. Hopefully flying within a few feet of it will be Eagle."

The two craft emerged right on cue and, thirteen minutes later, Eagle braked to a standstill and Collins reduced his speed to three inches a second while the nose probe of Columbia connected with the lunar module. There was a minor worry when Collins, thinking docking might not have happened correctly, tried to pull Columbia backwards. At that moment, a stabilising thruster fired on board Eagle, causing the two craft to wobble. Collins said: "That was a funny one. I went to retract and that's when all hell broke loose!"

Docking took place about three and a half hours after blast-off and Apollo 11 made another orbit while Armstrong and Aldrin removed any traces of lunar dust from themselves to reduce the risk of possible contamination before entering the command module.

Two hours and seven minutes after docking, Eagle was jettisoned and Columbia fired its engine to move into a higher orbit. As Eagle floated away, Collins remarked: "There she goes. It was a good one."

Apollo 11 continued to orbit at 5,355ft per second for another five hours before, fifty-six minutes after midnight and with Columbia on the far side of the Moon on its thirty-first orbit, the main engine was fired for two and a half minutes to boost the craft out of lunar orbit.

After another anxious period of radio silence, Armstrong told Capcom Charlie Duke: "Hey, Charlie boy, looking good here! That was a beautiful burn. They don't come any finer." And in a reference to the crew's spell in quarantine at the Lunar Receiving Laboratory, he added: "Time to open up the LRL doors, Charlie!"

SLEEPLESS NIGHT

NASA'S Medical Director gave the Moon men a clean bill of health tonight after their exertions on the lunar surface.

Doctor Charles Berry said he was delighted with Neil Armstrong and Buzz Aldrin's physical condition, despite their struggle to sleep in the lunar module.

The pair settled down to rest shortly after 5am and went for six hours and forty-seven minutes without any radio communications. But when Capcom Ron Evans called to wake them, Armstrong responded immediately. Evans asked: "How is the resting standing up there?" Aldrin explained: "Neil's been lying on the ascent engine cover. I curled up on the floor."

At a press conference tonight, Berry said medical data suggested they'd been dozing fitfully. He said: "I thought they would've been so tired that they would've needed more sleep than usual. Evidently this wasn't the case."

NASA officials were delighted with how the astronauts coped with walking in near-zero gravity. Doctor Robert Gilruth, Manned Space Flight Director, said: "They had more mobility than we expected. Their heart rates were in the nineties and they only used half the oxygen we expected. This might indicate that man will be able to wander further from his base than we might have hoped for." Gilruth, 55, added: "I'm hopeful that more ordinary physical specimens will be able to participate in space. I'm hopeful that even I may have a chance!"

GEOLOGY Q&A SESSION

The astronauts had a geology Q&A session today, adding to the excitement among scientists who will study the lunar samples.

Before leaving the Moon, Neil Armstrong and Buzz Aldrin answered a list of questions given to Mission Control by NASA geologists.

Asked how deep they were able to dig down to collect samples, Armstrong explained: "We got some from as much as three inches in the area where I was looking. In some other areas I'd get down just an inch or two and couldn't go any further."

An image taken from yesterday's live TV coverage of the Moonwalk. Before blasting off today, Neil Armstrong and Buzz Aldrin gave geologists further details of their observations.

Aldrin, who experienced difficulty hammering sample tubes into the ground, said: "I could get down to about the first two inches without much of a problem and then, as I'd pound it in about as hard as I could, it just wouldn't go much more than eight or nine inches. But even there it didn't, for some reason, want to stand up straight. It wouldn't penetrate in a way that would support it and keep it from falling over, if that makes any sense at all. It didn't to me."

Of soil extracted from under the surface, Aldrin said: "The material was a good bit darker. The way it adhered to the core tube gave me the distinct impression of being moist."

Describing the Moonscape, Armstrong added: "We're in a relatively clear crater field. The ground mass throughout the area is a very fine sand, to a silt. Immersed in this ground mass are a wide variety of rock shapes, sizes, textures, rounded and angular, many with varying consistencies."

The pair gave detailed descriptions of different rock compositions and Professor Gordon Goles, a geologist at the University of Oregon and one of the scientists at more than a hundred sites around the world who will study the samples, said he was intrigued by what he'd heard: "The images we saw weren't all that informative, but some of the observations and comments were very, very interesting. The descriptions began to sound very familiar to terrestrial geologists. The range of samples appears to be geologically diverse. It looks as though we can learn a lot. Now I'm just waiting to be part of it."

BASE STILL A MYSTERY

Space agency chiefs admit they still haven't pinpointed where Eagle landed.

At a press briefing tonight, Glynn Lunney, Flight Director, said NASA was certain the astronauts landed within the thirty-six square miles designated as 'Landing Site 2' and that technicians were confident they had narrowed down the location of Tranquility Base to within 3,000ft.

Lunney said the confusion had no bearing on take-off or rendezvous, but it was important to establish the location for the benefit of scientific studies.

Scientists in California, still without accurate co-ordinates, were again unsuccessful today as they fired laser beams at a reflector left on the Moon.

MANKIND 'IMMORTAL'

Landing on the Moon has "virtually ensured mankind of immortality", Doctor Wernher Von Braun claimed tonight.

Speaking at a press conference in Houston, the Saturn V mastermind said he had been one of the 'blue men' holding their breath in Mission Control as Eagle landed. He said: "I'm sure that, at that moment, I felt just as you did; so happy and so relieved they were safe. I remember thinking 'At last! We've finally made it!' I learned a long time ago to use the word 'impossible' with a great deal of caution. Back when we were just getting started, a lot of people said this day would never come.

"It's the culmination of a dream and the beginning of a very important new theme in the history of mankind. I think the ability for man to walk and actually live in other worlds has virtually ensured mankind of immortality. We can move to where we want to go, where other worlds can support our life or we can modify the environment."

Von Braun – who was at Mission Control for all the key moments – said Eagle would have crashed had it been under automated control for landing. "No doubt about it. Pilot skill saved the flight."

BIGGER THAN SHOWBIZ

People around the world were today still coming to terms with the enormity of the lunar adventure.

An early estimate from the European Broadcasting Union put the live worldwide TV audience for Neil Armstrong's first step on the Moon at more than 600 million. But this doesn't include hundreds of millions who watched delayed broadcasts in countries including the Soviet Union.

The historic event was the number one topic of conversation today, everywhere from major European cities to isolated villages in Africa.

In London, England, solicitor's clerk Anthony Crowhurst said he set an alarm for 2am UK time so he wouldn't miss the Moonwalk. He said: "It was absolutely, bloody marvellous! I leapt out of bed feeling pretty awful and just smoked away, watching until 4.45am. It was worth waiting for. I couldn't get over how calm those boys seemed to be."

Italian actress Gina Lollobrigida said: "Nothing in showbusiness will ever top what I saw last night." She clearly wasn't the only Italian watching. Police in Rome reported their quietest night in years, saying even the crooks had stayed at home to watch TV.

In Poland, Warsaw taxi driver Witold Borodzik invited neighbours who didn't have TV sets into his small apartment. He watched through the night and said: "It was one of the most thrilling moments of my life."

Canada's oldest licensed pilot, 83-year-old Tom Williams, from Sweaburg, Ontario, said: "I'm glad I lived to see it. It's astonishing to think what man has done since he first went into the air."

In Dodowa, Ghana, there are no TV sets and fewer than a hundred radios among the 3,000 people living in the small town twenty-five miles from the capital, Accra. Nagai Kassa VII, a regional chief, got up at 2am local time to listen and feared the astronauts might fall off the Moon. He said: "The Moon is so small, as I see it, that I didn't think there would be room enough for Eagle."

Annabel Parrott, a drugstore owner in St John, US Virgin Islands, said: "I didn't think it was possible and I certainly don't understand it at all. But then I don't understand TV, electricity or cameras either."

Australian Judge TJ Towns had a TV set brought into his courtroom in Wollongong and nominated a journalist as his 'TV controller for the day', ordering him to tell him when the Moonwalk started so he could adjourn proceedings. At one point, the judge even asked a witness to stand a little to one side so his view wasn't obscured.

There was relief in South America as coverage was unaffected. A satellite failure blacked out coverage of last week's launch, but the signal was successfully rerouted this time.

THE WORD ON THE STREET

Ordinary Americans had mixed views today on the Moon adventure, with reactions ranging from national pride to claims the whole thing was faked.

For some, beating the Soviets to the Moon was paramount. Leo Vigil, 51, from Albuquerque, New Mexico, said: "We've proved that we are number one." And Russell Caldwell, a 28-year-old insurance company mail clerk from Jacksonville, Florida, agreed: "I was proud to be watching. I was proud of them and I was proud to be an American."

Cathy Garcia, 17, from Conway, Pennsylvania, said the feat had given Americans "a renewed sense of patriotism". David Hildebrand, a 28-year-old executive assistant at a manufacturing firm in Miami, Florida, insisted: "For once, everybody in the world was thinking one good, common thought."

In Bethany, Oklahoma, Ron Hollis, 25, said: "We've got a baby girl and I hope that someday she'll be able to say she went to the Moon." Bill Parris, a 33-year-old furniture mover from Detroit, Michigan, admitted: "I've always hoped to see this in my lifetime and I kind of regret not getting my education so that perhaps I could've been part of the team on the ground."

The historical significance was the main benefit, according to many. Student Christine Wojtowicz, 20, from Depugh, New York, said it was "something to tell my children and grandchildren about", and Bob Barret, local Sheriff in Ellensburg, Washington State, called it "the greatest feat of the century". Charles Stuart, 23, from Portland, Oregon, marvelled: "When my grandmother was young, they were just learning how to fly. Now I wonder what's going to be happening by the time I'm her age."

Pioneering aviator Henry Dora, 76, from Bay City, Michigan, said: "When I first went up in a plane in 1912, everybody thought I was nuts. The highest I ever flew was 800ft. We never thought the United States would send a man to the Moon."

Lloyd Germyn, 19, from Eugene, Oregon, said: "I look up at the Moon now and I've got a completely different feeling. I really admire those astronauts. It's fabulous." Dismissing concerns about the cost, he added: "Americans spend more money on cigarettes and alcohol than on the space programme." Anthony Robertello, a bartender at Henry's restaurant in Schenectady, New York State, agreed: "It's worth the money. It's no more a waste of money than Columbus discovering America."

For many, the excitement was almost too much. Schoolteacher Barbara Guilford, 29, from Cheyenne, Wyoming, said: "I couldn't understand exactly what had transpired, but I was so exhilarated that I started laughing. I laughed for five minutes." Lorraine Javens, from New Brighton, Pennsylvania, said: "I couldn't sleep when I did go to bed. I'd been afraid that when they stepped out it would be like quicksand."

Photographer Pat Marchion, 62, from Fostoria, Ohio, was among those with mixed feelings about the space programme, saying: "This is the most wonderful thing I've ever seen. However, I don't believe they should go anywhere else. Why don't they take care of the Earth first?"

In New York City, Rose Rosen, 76, insisted: "It doesn't do any good. On this Earth there are so many people who are unhappy, forsaken and poor." And James Irvin, 25, from Boston, Massachusetts, complained: "It had no special significance to me. We have much more pressing problems here on Earth. If this is to the benefit of the masses, then why hasn't it helped the blacks in the ghettos?" Packing business owner Southern Jones, 62, from Detroit, agreed: "What benefit is it to humanity? Look at the poor people in the United States and yet they spend all these billions."

Others had more unusual reasons for objecting. Carolyn Baptist, a mother in her 20s from Baltimore, Maryland, said: "It all sounds pretty spooky to me. You know, like you read in those books. If there's something up there, they may bring it back. If they find some Moon men or something, they might chase the men back here." And 13-year-old Sarah Greer, from Rochester, New York, admitted: "Man, was I worried! I was afraid they might see somebody up there who wouldn't like the idea of our being up there."

Miller Taylor, 25, from Detroit, Michigan, tuned his TV to baseball instead. "You can't play baseball on the Moon," he said. "I don't understand what they're doing up there. At least I understand baseball."

Some people claimed the mission is faked. Barbara Hopkins-Day, 80, from McGehee, Arkansas, refused to watch, claiming it was a publicity stunt. She said: "I don't believe they've been there. If God intended for us to go to the Moon he'd have built a ladder up there." Ray Reppert, from Brighton Township in Pennsylvania, said: "When they got on to the surface, I said to my wife 'I wonder who's up there holding the cue cards?'"

SOVIET PROBE CRASH-LANDS

The Soviet Union's Moon mission came to a dramatic end today when Luna 15 apparently crash-landed.

Two hours before Neil Armstrong and Buzz Aldrin blasted off from the Moon, British experts monitoring the probe suddenly stopped receiving signals and, in an announcement later, *Tass* confirmed the mission had ended.

It is believed the craft came down in the Sea of Crises, about 500 miles from where Eagle landed. Data suggested Luna 15 was slowing down during descent but hit the surface at 300mph.

Sir Bernard Lovell, Director of Jodrell Bank Observatory in England, said his team lost contact with Luna 15 this morning and said they believed it crash-landed. In a typically-guarded statement, *Tass* said the mission ended at 11.47am and that Luna 15 had been a new kind of craft which had successfully demonstrated the ability to change orbital patterns, allowing control over where it landed.

After largely ignoring Apollo 11 yesterday, Soviet state TV today gave unprecedented coverage for an American mission. Moonwalk highlights were aired several times and Soviet media reports were unusually positive. Government newspaper *Izvestia* had a Moonwalk picture on page one and quoted cosmonaut Konstantin Feoktistov as saying: "What had been pure fantasy is now a reality. We rejoice at the success of the American astronauts."

A 'DOWNHILL RUN' NOW

The astronauts' families today began counting down the hours to their safe return.

With the blast-off from the Moon and the lunar orbit rendezvous negotiated, the sense of relief and confidence among the crew's loved ones was obvious.

After Neil Armstrong's wife, Jan, watched the lunar lift-off, the family's NASA liaison officer said: "I got the feeling that this, to Jan, was the real hurdle, that it looks to her like green and go from now on. She's more relaxed." Jan told reporters: "There's more danger in driving down a freeway than there is on these missions."

In Ohio, Armstrong's father, Stephen, was asked how he might react when he sees his son. He said: "I don't know what I'll say, but there won't be any tears in my eyes. There were plenty of tears before, particularly during the landing, but not anymore."

Michael Collins' wife, Pat, said: "I have no more worries from now to splashdown, but I've never really had any worries all along. It's very quiet in our house now. Everybody's sleepy. We didn't sleep much last night, but after the docking, everybody just sat back and relaxed."

Joan Aldrin clasped her hands above her head like a triumphant boxer as she spoke to reporters. She said: "It's all downhill now. I don't expect any more tense moments. I am, quote, thrilled, proud and happy, unquote."

Buzz Aldrin's sister, Faye Ann Potter, watched today's coverage with their father, Edwin Senior, in Brielle, New Jersey. Describing his reaction, she said: "Pleased? Absolutely! Relieved? Oh, yes!"

RIDDLE OVER FIRST WORDS

Neil Armstrong's first words when he stepped on to the Moon's surface have sparked a debate over whether he fluffed his lines.

A worldwide audience heard him say: "That's one small step for man, one giant leap for mankind." But some commentators have questioned whether he intended to say 'one small step for *a* man'. A NASA spokesman said: "I guess we'll just have to wait until he gets back."

CENSORED? HELL, NO!

NASA has denied censoring communications between the astronauts and Mission Control.

Describing a problem during docking, Michael Collins was heard to say: "I went to retract there and that's when all hell broke loose!" But when NASA issued the official transcript to the media, the quote became: "I went to retract there and then it all went no go."

NASA official EH Yeater, who supervises the typing pool that transcribes the communications, said: "It was an accident. If the transcription makes sense, we don't bother reviewing it. We just try to get it out as fast as we can."

CAMERA BOFFIN'S RELIEF

The man who designed the TV camera which filmed the Moonwalk has admitted he could hardly bear to watch in case it didn't work.

Stan Lebar, the electrical engineer who headed the Westinghouse Electric team who made the $250,000 camera, watched alone on a monitor in a cubicle at Mission Control. He said: "I worked four years for this. Just at that moment, waiting to see if it performed okay, I just couldn't bear to be with anybody else. How do I feel now? I haven't come down yet. Neil and Buzz and Mike may be on their way back, but I'm still up there with that great little camera."

The camera is set to spark a revolution in how TV shows and movies are filmed. It weighs only 7lb, has more than 250 miniaturised parts and runs on less electricity than a child's night light. Lebar said: "One thinks at once of live TV from the bottom of the sea, down coal mines; all sorts of unlikely places."

A separate TV camera was mounted inside the lunar module and the public will get to see footage of the landing itself – along with hundreds of still photos taken from inside Eagle and on the surface – once the astronauts return and the film has been held in quarantine. The still images issued to the press by NASA so far were taken from TV footage.

Armstrong and Aldrin had enough film with them to take 1,600 still photos and 1,680ft of movie footage. Michael Collins also had cameras on board Columbia.

'MOONDAY' HOLIDAY; FOR SOME

Millions of Americans enjoyed a day's holiday today to celebrate the landing, but many weren't so lucky.

Events in California were typical of the mixed response to the President's proclamation. Californian Governor Ronald Reagan backed the idea for a state-wide public holiday, but Los Angeles City Council voted that city employees should work as normal. Councilman Arthur Snyder said it would have cost LA $1.5 million in lost wages and insisted: "I can't bring myself to celebrate *that* hard."

The decision has infuriated the All-City Employees' Association, whose President, Alice Koumjian, has sent a telegram to NASA saying: "The city employees of Los Angeles have been refused a holiday in honour of the mission, but we want you to know that all city employees are with you 100 per cent."

Many private businesses chose not to close. They included M. Maggio Company of Philadelphia, which makes Italian cheese. Its President, Peter Maggio, said: "Moon or no Moon, we have to go on making cheese. After all, we can't turn the cows off."

For those people who had to work, there was only one topic of conversation. Kathy Bish, receptionist at a radio station in Toledo, Ohio, said: "The Moon landing was breathtaking. I stayed up after everybody else had gone to bed. Everybody at the radio station is starry-eyed today, either from staying up to watch the landing or from all the excitement of seeing it. Everyone felt they were right there with the men on the Moon."

COLLINS' HIGHLIGHTS REEL

Michael Collins will get a surprise present from NASA after splashdown – a specially-edited TV show of the mission highlights.

The Command Module Pilot had complained tongue-in-cheek that, as he didn't have a TV set, he missed the action.

Today, NASA revealed that clips of the mission – starting with the launch – are being flown to the USS *Hornet* recovery ship ready for Collins and his crewmates to watch while they are in quarantine.

Collins orbited the Moon alone for twenty-eight hours and, after the two craft docked today, Capcom Charlie Duke asked him: "How does it feel to have some company?" Collins replied: "Damn good, I'll tell you!" During today's NASA commentary, Douglas Ward, Public Affairs Officer, told the world: "Not since Adam has a human known such solitude as Mike Collins."

ORIGINAL ROCKET MAN

The widow of the world's first rocket pioneer says she cried after seeing his dream of a Moon landing come true.

Doctor Robert Goddard, an American engineer and inventor, who died in 1945 aged 63, is credited with paving the way for spaceflight after launching the first liquid-fuelled rocket in 1926. Goddard went on to launch rockets to heights of up to 1.6 miles and speeds of 550mph.

Speaking at her home in Worcester, Massachusetts, Esther Goddard said of the lunar adventure: "I cried. I was very happy and very sad, too. I wished Bob could've been here. He would've been beside himself with delight. That was his dream, sending a rocket to the Moon."

Goddard's work was initially ridiculed and Esther explained: "Oh, he had his setbacks. Those early days weren't easy, but Bob knew his detractors weren't qualified to judge him. He wouldn't give way to

Rocket pioneer Doctor Robert Goddard prepares to fire his liquid oxygen-gasoline rocket in 1926 in Auburn, Massachusetts.

anger or frustration. That was inside him, but few ever saw it. He knew that man would explore space, with the Moon only the first step. Bob believed, so unerringly, that some day all this would come about. He was certain that he was unlocking treasures of the universe."

The pioneer's 1926 breakthrough came in Auburn – six miles from Worcester – when his miniature liquid-fuelled rocket soared to 41 ft. His widow recalled: "We went to a field. Bob didn't want the rocket hitting any houses. We got there by car, bouncing over old farm roads with the rocket in a trailer.

"The rocket lifted up almost effortlessly. It was the most beautiful thing I'd seen in all my life. It came down in a cabbage patch. We all stood there for a moment, kind of dumbstruck. I can't recall that Bob said anything except maybe 'Let's go pick up the pieces'."

Recognising Goddard's importance to space exploration, Doctor Wernher Von Braun said: "He did most of the basic research and development that made rockets such as the Saturn V possible. His contributions are too numerous to recite."

MAN OF THE CENTURY

Ohio Governor James Rhodes today hailed Neil Armstrong as the "man of the twentieth century" as he officially launched plans to build a museum in his honour.

The $1 million museum, which Rhodes hopes will be built in Armstrong's hometown Wapakoneta within a year, will be funded through a combination of state taxes and donations.

Rhodes, who met Armstrong's father today to discuss the plan, said: "We want to establish this museum here before Neil's mementoes are lost to some institution in Washington or somewhere. Neil Armstrong will be the man of the twentieth century."

Armstrong is the latest history-making aviator to come from Ohio. The Wright brothers hailed from Dayton and astronaut John Glenn from New Concord. Rhodes added: "Neil's a son of Ohio soil and we're proud of him. Every time Ohio has a Wright brothers it takes us years to realise it."

DUKE'S DREAM

The mother of an astronaut whose voice has become familiar worldwide during the mission said today she hopes her son will get his own chance to walk on the Moon.

Charlie Duke's distinctive Southern drawl has become a feature of the mission, with him fulfilling Capcom duties during key moments, including the lunar landing.

Duke, 33, from South Carolina, was recruited by NASA in 1966, but has yet to make a spaceflight himself. He's widely tipped, though, to be the Lunar Module Pilot on a future Apollo mission, giving him the chance to walk on the Moon.

Today, his mother, Willie, said: "We're so proud he's had this part in it. He's had his glory just by working with such a crew. They're a most dedicated crew and a bunch of wonderful guys. I wish Charlie had been one of the men on the Moon this time, but I think he'll be involved in other flights."

FINAL SPLASHDOWN REHEARSAL

The recovery ship was declared ready for action today after holding a final full-scale rehearsal for Thursday's splashdown.

Every aspect of the operation was replicated down to the last detail, with the USS *Hornet*'s chaplain, Lieutenant Commander John Piirto, even playing the part of Richard Nixon.

Tonight, NASA recovery team chief Doctor Donald Stullken pronounced the exercise a total success, saying: "We're all set."

Captain Carl Sieberlich oversaw operations from the bridge as divers recovered colleagues playing the part of the three astronauts from a mock-up command module. The fake astronauts and the module were airlifted on board by helicopters.

FATHER OF THE LM

The man who masterminded the design of the lunar module spoke today of his delight at how well Eagle stood up to the demands of the mission.

Tom Kelly worked tirelessly for six years on the project after Grumman Aerospace won the main contract for the lunar module (LM) in 1962.

As chief designer, the 40-year-old has earned the nickname 'Father of the LM'. He was at Mission Control to watch the drama unfold and said: "It felt great; wonderful! It was a real culmination for a lot of people."

The former Air Force Lieutenant's involvement in the Moon shot effectively began in 1960 when Grumman appointed him project engineer for a manned spaceflight study. He said: "This gradually grew in scope and ultimately led to our winning the LM contract."

COLD WAR THAW

Hopes of a thawing of the Cold War were boosted today by a meeting between former US Vice President Hubert Humphrey and Soviet Premier Alexei Kosygin.

Humphrey was summoned to the Kremlin at the end of a two-week visit to the USSR and spent two hours in talks with Kosygin, who effectively shares power in Moscow with Communist Party First Secretary Leonid Brezhnev.

Humphrey said: "He was very complimentary about Apollo 11 and expressed his personal good wishes to the astronauts and their families. He gave me a message to take back to the United States. He wanted me to tell the President and the American people that the Soviet Union wants to work with the United States for peace."

CHAPTER 22

TUESDAY, 22 JULY
Launch Day Plus Six

"If it costs $100 billion to send a man to Mars, we simply don't have that kind of dough."
>   US Senate Democrat Leader Mike Mansfield

"No matter where you travel, it's always nice to get home."
>   Apollo 11 Commander Neil Armstrong

Charlie Duke (left) mans the Capcom desk at Mission Control with fellow astronauts Jim Lovell (centre) and Fred Haise. Duke was teased by the Apollo 11 crew today after mistaking the Moon for the Earth.

SPACE TV COMEDY

An embarrassing live TV blunder by Charlie Duke was the highlight today for the astronauts as they coasted homeward.

They teased Capcom Duke after he mistook the Moon for the Earth during a live telecast. As the show began, Duke said: "The focus is a little bit out. We see Earth in the centre of the screen." After a brief silence, Buzz Aldrin replied: "I believe that's where we just came from!"

A flustered Duke responded: "Is it, huh? Hey, you're right!" Aldrin told him: "It's bad enough not finding the right landing spot. You haven't even got the right planet!" Duke admitted: "I'll never live that one down."

When Michael Collins pointed out he should be able to tell which one was the Moon because it would get smaller as they got further away, Duke said: "Aww, that's enough guys!" Collins ribbed Duke over the fact he might have "fun at the press conference after this shift", and Duke joked: "I'm going to sneak off through the back way."

In another offbeat moment, communications were interrupted by shrieking noises and eerie laughter. Duke asked the crew: "You sure you don't have anybody else in there with you? We had some strange noises on the downlink and it sounded like you had some friends up there." The astronauts insisted they had no explanation, but NASA sources said they thought the crew were playing a prank on Duke.

The exchanges summed up the relaxed mood after the drama of the previous few days. At 1.39pm the spacecraft crossed the point where lunar gravity ceased to have a dominant influence. The pull of the Earth will steadily increase Apollo's speed from 2,720mph to the 24,760mph it will be travelling at when it plunges back into the atmosphere.

Tonight's telecast lasted twenty minutes and ended with a dramatic shot of the Earth 175,000 miles away, and Neil Armstrong signed off by saying: "No matter where you travel, it's always nice to get home." Asked for the forecast for splashdown, Duke said: "It looks like it's going to be real nice. Navy called up calm seas for you guys."

COAST-TO-COAST PARTY

Plans for a spectacular day of celebrations to salute the Moon heroes were revealed today.

The White House says three parades are planned for 13 August, just forty-eight hours after the crew leave quarantine. In one hectic day, the astronauts will make a coast-to-coast tour, travelling on the President's plane, Air Force One, from Houston to New York, on to Chicago and then Los Angeles.

Ticker-tape parades will be held in New York and Chicago before a dinner is hosted by Richard Nixon for 1,500 guests at LA's Century Plaza Hotel.

Neil Armstrong's hometown will hold a parade in August, it was confirmed today. Charles Brading, head of the planning committee in Wapakoneta, Ohio, said: "I expect thousands will come, maybe even hundreds of thousands."

NIXON: WE'LL FIND LIFE IN SPACE

Richard Nixon today predicted man will find life on other planets by the year 2000.

The President was in a buoyant mood as he attended two public functions before setting off tonight to watch splashdown. Speaking to 2,000 overseas students on the South Lawn of the White House, he said: "This is no time for the pessimists. This is the time for the optimists, the idealists. In the year 2000 we on this Earth will have visited new worlds where there will be a form of life."

He told them it was "a shame" that people in the Soviet Union and China hadn't seen the lunar drama live. "I hope that when the next great adventure into space takes place that it will be one in which Americans will be joined by representatives of other countries so that we can go together. I want the time to come when the Chinese people and the Russian people and all the people of the world can walk together and talk together."

Nixon also entertained 500 guests at a reception to coincide with baseball's All-Star Game. Guests included players, coaches and team owners, plus the likes of baseball legend Joe DiMaggio and the widow of possibly the greatest player ever, Babe Ruth. Nixon planned to attend tonight's game, but torrential rain forced it to be postponed until tomorrow.

NERVE GAS SCANDAL

America today admitted responsibility for a leak of deadly nerve gas in Japan, casting a political shadow over Richard Nixon's goodwill tour of Asia.

As Nixon left Washington tonight, Vice President Spiro Agnew said: "The message of the olive branch left on the Moon by astronauts

Armstrong and Aldrin will be carried by the President in his personal effort to find the road to peace, progress and human understanding for all mankind."

But that statement had a hollow ring after the Pentagon confirmed the USA had secret stocks of chemical weapons on the Japanese island of Okinawa, where it has military bases under a post-World War II treaty, and that these were the source of a leak on Friday which left twenty-four Americans needing emergency treatment.

The Pentagon has confirmed "certain chemical munitions, including some toxic agents" will be removed, but Japan has expressed its anger and Nixon faces a backlash on his tour. In the Philippines, which will be his first stop, Foreign Secretary Carlos Romulo said there was "deep concern" that such weapons might be stored elsewhere in Asia.

The President flew from Washington to San Francisco tonight and tomorrow will travel 3,240 miles across the Pacific ready for splashdown.

EXPERIMENT DOOMED

Scientists admitted today that a multi-million-dollar experiment set up on the Moon is likely to stop working within days.

The seismometer – designed to pick up lunar tremors – is overheating and has already reached 215 degrees Fahrenheit. That's forty-five degrees above what it is designed to withstand.

The device has worked well so far, but NASA believes rocket exhaust may have charred the seismometer, meaning it would absorb heat from the Sun. In four days, the Sun will be directly over the landing site and Jack Small, Lunar Experiments Co-ordinator, admitted: "My best educated guess would be that it won't survive."

However, Doctor Garry Latham, of Lamont Geophysical Observatory at Columbia University, New York, insisted: "We're very pleased with this experiment. We're receiving data from an extra-terrestrial body, which is, in itself, an accomplishment."

The seismometer setback follows the continuing failure of scientists to hit a lunar reflector with laser beams. Doctor James Faller, Co-director of the Lick Observatory, California, said: "Since we don't know the exact location, all we can do is shoot at several locations. If we find it soon, it'll be pure and simple luck."

Doctor Garry Latham (left) studies data being received today from a seismometer set up on the Moon.

PERILOUS LANDINGS PLANNED

America's astronauts will face increasingly dangerous landings under plans outlined today for future Apollo missions.

NASA confirmed that the Ocean of Storms is the preferred landing site for Apollo 12, which will be launched in November or December. Flights made between now and 1973 will all land at different sites. The Ocean of Storms, like the Sea of Tranquility, is a 'mare' – a relatively flat plain – but more hazardous sites are earmarked for later missions.

Doctor Ted Foss, Chief of the Geology Branch in Houston, said it would be necessary to take greater risks. He said: "If we're really eventually to come to a position where we understand the general history of the Moon, we can't be satisfied with just landing on mare sites."

Outlining the plans, Doctor George Mueller, Associate Administrator for Manned Space Flight, said increasingly complex scientific experiments would be set up on the Moon and that later landings would see the lunar module modified to transport the proposed Lunar Roving Vehicle, allowing crews to venture further.

No plans have been drawn up to land on the far side of the Moon. Jim Sasser, Director of the Lunar Mapping Laboratory, said: "That could still be quite a few years away. Whenever our spacecraft go behind the Moon we're no longer able to see them on TV or talk to them. There is a simple answer to that. We could put around the Moon a number of communications satellites, just as we have around the Earth, and we'd have constant sight and sound of any crew landed on the back of the Moon."

NASA also revealed details of its proposal for the world's first space station. The 'workshop' would be launched in 1972 and would orbit 250 miles above the Earth. Astronauts would stay there for as long as fifty-six days at a time.

A Saturn V rocket would be used to orbit a station equipped with an experimental device called the Apollo Telescope Mount. A three-man crew would be launched twenty-four hours later by a smaller Saturn 1B rocket to rendezvous and dock with the station. Scientists hope the telescope, which would allow astronauts to study the Sun and stars from above the veil of the Earth's atmosphere, could help establish the origins of the universe.

NUCLEAR MARS FLIGHT PLAN

Work is already under way to design a nuclear-powered engine capable of taking men to Mars in the 1980s, it emerged today.

NASA and the Atomic Energy Commission are pumping an initial $1.5 billion into research by Pennsylvania-based Westinghouse Electric aimed at harnessing the same energy used by a hydrogen bomb.

Doctor George Mueller, NASA Associate Administrator for Manned Space Flight, said: "We need to develop a new source of energy, a new source of giant thrust to carry spaceships from this solar system to its neighbours in the galaxy. We do know, in principle, what source of energy it could be. It's the same one that fuels our Sun; the fusion of hydrogen to make helium. All it requires is an invention to harness it."

Doctor Walter Esselman, Westinghouse astro-nuclear laboratory chief, said the company had been working on inventing a nuclear engine since 1954. He said: "The engine will be about the size of a desk and will have the power of the Hoover Dam. If this engine were used in the Moon landing, the astronauts could have collected an extra 25,000lb of rock and stayed on the Moon for 100 days."

Esselman said a Mars mission could happen "in the 1980s" but conceded the funding needed was "a question of national goals".

Whether all of NASA's future Moon missions – let alone a manned flight to Mars – will happen depends on continued funding. Mike Mansfield, Democratic Leader in the Senate, said: "If it costs $100 billion to send a man to Mars, we simply don't have that kind of dough."

But George Miller, Democrat Representative for California and House Science and Aeronautics Committee Chairman since 1961, hit back: "Politicians are smart enough to know you can't cut an American dream in half. We've gone to the Moon. You can't just suddenly say we'll stop there."

Writing in today's *New York Times*, John Glenn said astronauts' "very existence" relied on "the say-so of the public" and that winning over national opinion was vital if funding was to be secured for future exploration. Of the successes so far, he said: "For us, as astronauts, to aspire, the United States had to aspire first."

STAFF IN LOCKDOWN

A dozen Apollo 11 backroom staff are already in quarantine in Houston ready for the astronauts' return.

The team – including medical and research technicians, a photographer and a public information officer – went into lockdown at the Lunar Receiving Laboratory (LRL) on Sunday just before the Moonwalk.

They will help with medical tests, the mission debrief and general housekeeping. They will be joined on Thursday by a doctor and a scientist, who will arrive in a special quarantine van with the astronauts. The building has fifty-eight reception rooms and could house as many as 120 people. Facilities include sleeping quarters, a dining room and kitchen facilities, as well as numerous laboratories.

NASA officials have again played down concerns over the danger of 'lunar bugs' and defended the precautionary measures. Doctor William Kemmerer, Chief of Preventative Medicine, said: "Although the probability of harmful bugs is very, very low, the value of what must be protected – all the people on Earth – is very, very high. The programme we have mounted is what one might call an insurance programme. We're protecting that which is at risk; the terrestrial biosphere, the totality of life here on Earth."

Doctor Elbert King, LRL Curator, said the danger of harmful bugs was "lower than your chance of being hit by a meteorite on Earth". He believes any harmful organisms would have arrived on Earth many years ago due to "a constant interchange of materials between Earth and the Moon". He added: "I work with meteorites all the time and I don't break out with purple crud."

RUB-A-DUB-DUB ...

The Apollo 11 heroes are set for an undignified return to Earth.

Quarantine procedures mean the three astronauts and a Navy diver will spend twenty minutes squirming about in a rubber raft scrubbing each other down from head to toe with disinfectant.

The bizarre scene will be played out once the crew have changed into biological suits handed to them by Lieutenant Clancey Hatleberg.

John Stonesifer, Associate Recovery Team Leader on board the USS *Hornet*, said: "It'll be quite a scene; rub-a-dub-dub, four men in a tub!

234

It's ironic, but when millions of people all over the world get their initial glimpse of the astronauts on their return, they'll knock themselves out laughing!"

SOVIETS 'NO LONGER MASTERS'

America's new domination of the space race will force the Soviet Union to co-operate in future missions, it was predicted today.

The crash-landing on the Moon of the unmanned Luna 15 probe on the same day that Neil Armstrong and Buzz Aldrin were on the lunar surface has emphasised how far behind the Soviets have fallen.

Sir Bernard Lovell, Director of Jodrell Bank Observatory in England, said today: "After Apollo 11, the American nation can now do almost anything if they wanted to. This is the first time the United States has been demonstrably superior. American approaches for collaboration may be received with sympathy in the Soviet Union as they can no longer regard themselves as masters in space."

LETTERS FLOOD IN

Space agency office staff are working flat out to deal with tens of thousands of letters from around the world.

NASA replied to more than 100,000 letters in 1968 and expects that number to be comfortably beaten this year thanks to Apollo 11. Charles Biggs, Curator of the Exhibits and Artefacts Office in Houston, said: "It really is amazing. Letters are coming in from all over the world."

Most correspondents either offer their congratulations or ask for souvenirs, such as signed photos of the astronauts. But there are some more unusual requests. Biggs said a man from Brookline, Massachusetts, wrote asking: "I've been promising my wife a piece of the Moon for many years. You'd be doing humanity and romance a wonderful deed if you sent her a piece of the Moon, no matter how small."

A woman from Chicago, Illinois, wanted to "lay lunar material" on sick people to "regenerate" their failing organs, and someone calling themselves the "Minister Plenipotentiary of the Democratic People's Republic of Syrtis Major" demanded NASA stay away from the Moon. The letter warned: "Tranquility is the property of my government. The landing is a violation of territorial privileges."

Gail Martens, who heads a team of six that deals with replying to letters, said they sent 100,000 responses last year. She said: "We were almost caught up with it all. Then Apollo 11 came and the roof fell in!"

Although most letters are positive, she recalled one from a Dallas woman who was upset about the space jargon used during missions. The woman complained: "It's roger, roger, roger, roger, roger, roger all the time. This roger mess is silly, ridiculous and asinine."

One woman from Richmond, Virginia, wrote asking for an astronaut application form, saying: "You may need a man who can stand the altitude. If so, I'd like to volunteer my husband. He comes home high as a kite every night."

RE-ENTRY SIGHTSEERS

Passengers on an Australian commercial flight are set to see Apollo 11 hurtling back through the Earth's atmosphere.

Qantas has put back the departure of Thursday's Sydney-to-Honolulu service after its pilot, Captain Francis Brown, contacted NASA for the precise timings and location for re-entry and realised passengers would have a unique chance to see the spacecraft.

Qantas has renamed the flight a 'Lunar Special Service' and says some seats are still available. Brown said the aircraft should be parallel to the spacecraft, about 600 miles away, near the Pacific islands of Gilbert and Ellis. He believes that for about four minutes people will be able to see a trail of "bright explosions" as the command and service modules separate and the craft emerges.

WIVES' PARTY NIGHT

Jan Armstrong, Joan Aldrin and Pat Collins were the guests of honour at a pre-splashdown party in Houston tonight.

Apollo 8 astronaut Bill Anders, a member of the back-up crew for the current mission, hosted the bash. While the party was in full swing, Michael Collins asked Mission Control: "Just wondering how everything's going on the home front? All the wives and kids in one piece?"

Capcom Charlie Duke told him: "All the gals are having a little party tonight as far as I know." Collins responded: "Oh good, glad to hear it!"

This afternoon, George Mueller, NASA Associate Administrator for Manned Space Flight, visited each of the wives to thank them for their co-operation before and during the mission.

SPACE CITY WILL GO WILD

Houston is bracing itself for the mother of all splashdown parties.

NASA Highway 1, the main road through the Texan space community, has been the hub of wild celebrations after previous missions, with the scenes which greeted the return of Apollo 8 likened to a cross between Mardi Gras and 4 July.

Hotel and bar bosses admit they're expecting a crazy night on Thursday. Doyle Graham, Manager of the Sheraton Kings Inn, said: "This could be a lulu! There's nothing like a splashdown party here. You've got a greater concentration of brainpower here during a spaceflight than anywhere in the world. These people have been working under tremendous pressure and strain. They're not a bunch of drunks, but when they let their hair down …"

Jess Young, Nassau Bay Police Chief, is confident it won't get out of hand. He said: "Oh, they'll have a helluva crowd, but they've been fairly orderly in the past. I say *fairly* orderly. I hope everybody has a ball. I think they will. In fact, I'm *sure* they will."

ADVICE ON FIRST WORDS

Neil Armstrong's sister has revealed how he was inundated with letters suggesting what his first words on the lunar surface should be.

June Hoffman – who visited Neil's wife, Jan, in Houston today – said: "I know full well that Neil was given a great deal of suggestions. In fact, the mail was stacked up."

She said the only time she's been worried during the mission was in the last few minutes of the descent to the surface. She admitted: "That was when I crossed my fingers the tightest."

CHAPTER 23

WEDNESDAY, 23 JULY
Launch Day Plus Seven

"The weather is clobbering in at our landing point. We don't want to tangle with one of those thunderstorms."
Capsule Communicator Charlie Duke

"Space, news and politics. What else is there?"
TV News Presenter Walter Cronkite

US President Richard Nixon is welcomed on board the USS *Hornet* recovery ship by a guard of honour.

STORMS FORCE SPLASHDOWN SWITCH

Thunderstorms in the Pacific tonight forced NASA to change its splashdown target zone.

Mission Control informed the astronauts shortly before they settled down for their final night's sleep in space that they'll have to land 247 miles closer to Hawaii than planned. Capcom Charlie Duke said: "The weather's clobbering in at our targeted landing point due to scattered thunderstorms. We don't want to tangle with one of those."

The news came at the end of an uneventful day during which the spacecraft gradually picked up speed as Earth's gravity continued to pull it home. At 3.58pm the craft reached the halfway point when it passed the 118,321-mile mark. The crew were so quiet that Bruce McCandless radioed to ask at one point during his Capcom shift: "Apollo 11. Are you still up there?"

Tomorrow the crew will face their final moments of danger when, at about 12.30pm, the command module will turn into a fireball as it plunges into the atmosphere. Columbia will hit the atmosphere at an altitude of 400,000ft and it will take around fourteen minutes from that point until splashdown. It will fly blunt end first, as that part of the craft is heavily shielded to withstand temperatures of up to 5,000 degrees Fahrenheit caused by the friction.

The craft will gradually slow and, once it emerges from the atmosphere, a parachute will be deployed when Columbia is at an altitude of 23,300ft and travelling at 306mph.

The craft must hit a thirty-mile 'window' in the atmosphere. If it doesn't, the crew face certain death. Too sharp an angle of descent would see the craft break up. Too shallow an angle would cause it to skip off the atmosphere like a stone skimming on the water and bounce back into orbit.

CREW THANK UNSUNG HEROES

The astronauts tonight paid tribute to the hundreds of thousands of unsung heroes behind the Moon mission as they made an emotional final telecast.

Passing the hand-held camera between them, each man spent several minutes reflecting on the historic flight. Michael Collins thanked those who had put "blood, sweat and tears" into making it a success. He said: "This trip to the Moon may have looked simple and easy. I'd like to assure you this hasn't been the case. All you see is us, but below the surface are thousands and thousands of others. To all of those, I'd like to say 'Thank you very much'."

Buzz Aldrin said: "We've come to the conclusion that this has been far more than three men on a voyage to the Moon, more still than the efforts of a government and industry team, more even than the efforts of one nation. We feel this stands as a symbol of the insatiable curiosity of all mankind to explore the unknown."

Neil Armstrong thanked the "giants of science" who made the mission possible. Signing off, a clearly emotional Armstrong said: "We'd like to give all the Americans who built those spacecraft, who did the construction, design, tests and put their hearts and all their abilities into those craft, to those people tonight, we give a special 'Thank you'. To all the other people listening and watching, God bless you and good night from Apollo 11."

The telecast ended with a dramatic shot of the Earth. Joan Aldrin watched at Mission Control and, in a dig at critics who said the astronauts were too business-like, she said: "I should hope those guys have proven to the world how articulate they can be."

LANDING WAS ALMOST ABORTED

NASA officials have admitted how close the lunar landing came to being aborted.

The lunar module was almost out of fuel by the time Neil Armstrong set it down and the landing was also threatened by a series of warning alarms. Lieutenant General Sam Phillips, Apollo Programme Director, conceded: "We damn near didn't make it. Neil might see it differently, but I think we came awfully close to having to abort."

Phillips said the landing was saved by "the most magnificent team performance I've ever seen in my life", and he believes the hours Armstrong spent piloting the vertical take-off Lunar Landing Research Vehicle proved vital. The astronaut cheated death last year when he parachuted to safety moments before it crashed and exploded.

Phillips admitted NASA had reservations about the vehicle, but said: "We decided the risk involved in flying it was more than offset by the worth of the training. Neil's ability to make the lunar module move away from the crater was a function of his high state of training. It really paid off."

The landing was almost called off when a series of alarms were triggered in the cockpit. Stephen Bales, a 27-year-old Guidance Officer, relied on his in-depth knowledge of the spacecraft systems to tell the astronauts to ignore the alarms. He realised it was just the on-board computer's way of warning it was struggling to cope with so many tasks at once.

Doctor Chris Kraft, Director of Flight Operations, admitted: "The computer was right on the ragged edge." Reflecting on the tension and emotion in those crucial moments, George Hage, Mission Director, said: "Bales and some of those other guys ought to get medals. There were some wet eyes in that control room."

NIXON CALLS WIVES

Richard Nixon said a personal thanks to the astronauts' wives today after a last-minute scramble to find them.

The President phoned from San Francisco, but the surprise calls ran into a hitch because all three women were attending a lunch at the Lakewood Yacht Club in Houston.

Pat Collins received her call just before going out, but the other two wives had already left home when Nixon phoned. Eventually the President spoke to them at the yacht club. Joan Aldrin said: "It came as such a surprise. It was a very satisfying conversation."

Later, the President and First Lady flew across the Pacific to Johnston Island. Pat Nixon won't be joining her husband on board the USS *Hornet* for splashdown. She'll fly straight on to Honolulu.

She said she would be "glued to the TV again" during splashdown and said of her emotions during the lunar landing and take-off: "All of us had our hearts in our throats." The First Lady said she hoped the mission would have a positive impact on American society. "I think this is going to draw us closer together. In San Francisco, I even saw a couple of hippies who were very, very happy about it. Maybe they won't be hippies very long; that's my hope."

MOONQUAKE ARGUMENT

Scientists in charge of the Moon experiments made their first preliminary findings public today and are already arguing about what the results mean.

A dozen experts held a press conference in Houston to discuss data received from a seismometer left on the surface, but there were disputes over an "exciting" event already detected by the equipment.

They revealed that, at 4.20pm yesterday, data indicated a large shockwave. While some of the scientists believe it was a moonquake, others insist it was caused by a meteor strike. Earthquakes are caused by the stresses and strains of a hot, active core. If there are moonquakes, this would indicate the Moon has a similar composition to the Earth.

The scientists seem to agree that the Moon has layers, because the shockwave appeared to travel through a crusty material overlaying a more solid interior. However, the experts are divided over the cause of the shockwave.

Some compared the data as similar to an earthquake registering five on the Richter Scale. One scientist even said he'd be claiming a case of Champagne from a wager he'd made with another expert that moonquakes do exist.

But geologist Doctor Maurice Ewing claimed his colleague "may have an argument on his hands when he tries to collect his case of Champagne". He said that, while he was "tremendously excited", it was "too early in the game" to say with assurance that it was a moonquake.

HUNDREDS SCREENED FOR BUGS

A specialist in rare communicable diseases arrived in Houston today to help NASA doctors carry out an unprecedented screening programme for potential 'Moon bugs'.

Hundreds of Manned Spacecraft Center staff are being monitored under a special programme covering anyone who might conceivably be contaminated by lunar organisms.

The National Communicable Disease Center in Atlanta has sent one of its doctors to assist. Blood samples have been taken from staff and they have been ordered to report even the slightest sign of illness. Doctor George Brooks, of the Communicable Disease Center's Epidemic Intelligence Service, said: "The purpose is to stay one step ahead of anything that might happen."

SOVIETS HINT AT CO-OPERATION

A leading Soviet space scientist today gave one of the clearest hints yet that they're ready to work together with America.

Writing in the Communist Party newspaper *Pravda*, Leonid Sedov, a member of the Soviet Academy of Science, said one country alone couldn't afford "the huge material expenditure and concentration of countless highly-qualified workers and specialists" to realise future projects.

Although the USSR has lost the race to the Moon, Sedov insisted: "Soviet space research will continue in various directions, including in the sphere of manned spacecraft. Human feelings and observations can't be completely replaced by automatic stations."

Astronaut Frank Borman said: "The Russians have had the opportunity to co-operate. It's nice to talk. It'd be nice if we got a little action besides talk."

STUDENTS 'ALIENATED'

An Ivy League university academic today claimed the lunar landing has further alienated young Americans rather than helping to heal the cracks in US society.

Students have been at the heart of protests over civil rights and Vietnam during a troubled end to the decade. Asked whether the Moon mission had opened new horizons for his students, Doctor George Wald, a biochemist at Harvard University in Massachusetts, said: "I'm afraid they see in this an exercise in great wealth and power, heavy with military and political overtones. They feel a little more trapped, a little more disillusioned, a little more desperate.

"What should have been a great flight of the human spirit comes to us heavy with threat. Those almost miraculous guidance systems that so uncannily find their targets, will they one day be guiding missiles to find *us*?"

'THE OTHER ASTRONAUT'

A veteran TV presenter dubbed 'The Other Astronaut' for his love of the space race explained today how he coped with his marathon stints on air during the mission.

Newsman Walter Cronkite is so dedicated to learning about space exploration that last year he even tried out a reduced gravity simulator used in training by astronauts.

Walter Cronkite has been credited by industry experts as the key to CBS's stunning Apollo 11 ratings victory. Figures released today show it has claimed forty-five per cent of the US audience to NBC's thirty-four and ABC's sixteen.

Cronkite has been with CBS since 1950 and has covered the space race from the outset, with his mix of expert knowledge and enthusiasm making him a household favourite. He shed a tear live on air when Eagle landed. When the astronauts safely blasted off again, he exclaimed: "Oh boy! Hot diggity dog!"

The 52-year-old newsman was on set from 9.45am on Sunday until 3.15am on Monday. He grabbed a few hours' sleep before another seventeen-hour shift. Cronkite said he had to stay on air because of the mission's significance. Asked whether he felt tired, he said: "Yes, but you don't think of that. This was something we've been aiming for all these years. We've been building toward this. The only thing comparable to it was splitting the atom, but we couldn't cover that. It was done in secret.

"I grabbed a bite every time one of our three-minute film pieces was on. I drank a lot of tea and soft drinks. They kept me going. I almost got caught out once. I'd just taken a bite of a sandwich when the camera came back to me. I got it over to my left cheek just in time and turned my head!"

He also did a seventeen-hour stint during the US election in November and says there's nowhere he'd rather be during major events. "Space, news and politics. What else is there?"

Cronkite said astronaut Wally Schirra has played a major part in the CBS success. "Thank God for Wally Schirra," he said. "He had such confidence. It was contagious. He was a wonderful crutch."

CHAPTER 24

THURSDAY, 24 JULY
Splashdown Day

"Many of us still can't believe that the goal we set out to achieve in 1961 has been achieved."
  Apollo Spacecraft Programme Manager George Low

"We congratulate you, courageous astronauts. The glorious dream of visionaries and scientists has come true."
  Moscow Radio News Flash

US President Richard Nixon chats to the astronauts through a window of the Mobile Quarantine Facility on board the USS *Hornet*.

'GREATEST WEEK SINCE CREATION'

Richard Nixon welcomed the astronauts home today and told them: "This is the greatest week in the history of the world since creation."

The President was on board the USS *Hornet* recovery ship as the spacecraft plunged into the Pacific at 12.50pm. Apollo 11 safely negotiated the final perilous phase of the mission, withstanding temperatures of up to 5,000 degrees Fahrenheit as it hurtled through the atmosphere.

The historic flight lasted 195 hours, 17 minutes, 49 seconds from the moment of launch from Cape Kennedy to landing 950 miles south-west of Hawaii. The craft came down within three miles of the target zone and the mission lasted just twenty-seven seconds longer than NASA predicted.

After Neil Armstrong, Buzz Aldrin and Michael Collins were safely airlifted on to the ship, Nixon spoke to them via an intercom through a window of the Mobile Quarantine Facility and told them: "Gee, you look great!"

Nixon said: "As I travel to Asia and Europe, I'm going to find that, as a result of what you've done, the world is closer together. We can reach for the stars as you have. I'm the luckiest man in the world. Not only because I have the honour to be President, but particularly because I have the privilege of speaking for so many in welcoming you back to Earth."

Nixon told them he had phoned "three of the greatest and most courageous ladies in the whole world", and that "from Jan and Joan and Pat I bring their love and their congratulations".

The only minor hitch with splashdown came when Columbia turned upside down after hitting the water. It was quickly righted by flotation bags. Nixon asked: "As we saw you bouncing around out there, I wonder if that wasn't the hardest part of the journey? Did any of you get seasick?" Armstrong replied: "No, we didn't. It was one of the hardest parts, but it was one of the most pleasant, we can assure you."

The astronauts, who spoke to their wives by phone tonight, won't give a press conference until they are released from quarantine.

NO SIGN OF 'MOON BUGS'

The astronauts were given a clean bill of health this afternoon after being whisked into quarantine.

Doctor William Carpentier – who will be in lockdown with the Moon men until 11 August – carried out a medical check within an

The astronauts, having changed into biological suits and breathing masks after today's splashdown, wait in a rubber raft as diver Lieutenant Clancey Hatleberg closes the hatch of the command module.

hour of splashdown. He said: "There's no sign of infection of any kind. This is only a cursory, preliminary examination, but I believe they took absolutely no harm from their Moon landing. They're in great shape."

Today's finale began when Mission Control woke the astronauts five and a half hours before splashdown and informed them that a planned course correction wasn't required. Columbia was 47,000 miles away and hurtling towards Earth at a velocity of 10,000ft per second. This increased to around 36,000 at re-entry.

Eight minutes before re-entry the service module was jettisoned and, just before radio contact was lost when the cone-shaped command module breached the atmosphere, Neil Armstrong told Mission Control simply: "See you later."

When the spacecraft emerged from the atmosphere, having already slowed during re-entry, its parachutes opened as planned and it drifted into the ocean at only 21mph. Richard Nixon, watching through binoculars from the USS *Hornet*, was among those to catch a fleeting glimpse of the spacecraft during its fiery re-entry. But, with Columbia coming down in cloud cover fourteen miles from the recovery ship, he and TV viewers worldwide didn't see splashdown.

Although recovery aircraft reported spotting the spacecraft during its descent, there was tension in Houston as the communications blackout with Columbia – expected to last three minutes and fifty-three seconds – went on for nine minutes. Capcom Ron Evans made nine attempts to make contact before Armstrong finally answered. As the recovery team swung into action, Collins reported: "We're just fine. Take your time."

Stringent protocol for disinfecting the astronauts and spacecraft were carried out to the letter by Navy divers before Armstrong, Aldrin and Collins – wearing biological suits, hoods and breathing masks – were picked up from a rubber raft and airlifted on to the ship. They waved briefly as they left the helicopter and made the short walk to the Mobile Quarantine Facility (MQF). By 2pm they had undergone a medical and, at 2.55pm, were speaking to the President. An hour later Columbia had been airlifted on board and placed in quarantine.

Columbia was attached to the MQF by a plastic tunnel and Doctor John Hirasaki, a NASA scientist who has been waiting inside the facility since Saturday, set about removing the lunar sample boxes, photographic and movie film, data tapes and other items. He shut down the capsule's systems before passing the sample boxes out via a decontamination lock ready for them to be flown to Johnston Island and then on to Houston.

SHARKS SCARED OFF

One of the divers involved in today's operation has revealed how the splashdown site was swarming with sharks.

Rehearsals during the last fortnight were repeatedly blighted by schools of sharks and Lieutenant Clancey Hatleberg said sharks "were all around" during today's recovery, but that the noise of the helicopters hovering overhead put them off coming too close.

Hatleberg and the astronauts were all wearing biological suits, hoods and breathing masks while they were in close contact today, but the diver was aware he'd have been quarantined if something had gone wrong. He joked: "My wife warned me that if I had to spend twenty-one days locked up in isolation she'd send her lawyer to see me!"

The *Hornet* team completed the recovery operation in three hours and three minutes. That was faster than any of the eighteen rehearsals. Commander Donald Jones – the helicopter pilot who airlifted the astronauts to the ship – said: "It was all very easy."

Some of the ship's crew caught a glimpse of Columbia during re-entry. One sailor, Kim Bunderson, 21, from Ogden, Utah, said: "I saw it for a few seconds, like a string of fire."

PRICELESS CARGO

Space agency chiefs were taking no chances today as the lunar samples were flown to Houston.

The two boxes are being transported across the Pacific in separate Air Force jets in case one should crash. Doctor Thomas Paine, NASA Administrator, said the samples were worth "more than all the gold we have at Fort Knox".

Experts will examine the samples once they've undergone quarantine precautions at the Lunar Receiving Laboratory. These will include exposing mice, insects and plants to the lunar rocks and dust. Doctor Gene Simmons, from the Institute of Technology in Massachusetts, said: "No other specimens in the entire history of science will have had such a large number of measurements made on them or such a large number of very high-calibre scientists working on them." And Doctor Elbert King, Curator of the Houston lab, said: "I won't be surprised if these rocks start a lot of arguments."

MISSION ACCOMPLISHED

Space agency chiefs admitted to a sense of relief and disbelief today at hitting John F. Kennedy's Moon deadline.

When the mission officially ended today, a giant screen at Mission Control displayed the words from Kennedy's 1961 Moon speech alongside the message: "Task accomplished … July 1969."

George Low, Apollo Programme Manager, said: "Many of us still can't believe that the goal we set out to achieve in 1961 has been achieved."

Doctor Robert Gilruth, Director of the Houston centre, agreed: "We have a feeling of elation and relief. We're supremely happy."

Apollo Programme Director, Lieutenant General Sam Phillips, said the hundreds of thousands of Americans who made it possible were "the strongest team ever assembled in the history of man".

KING OF ROCKET CITY

Huntsville lauded adopted son Doctor Wernher Von Braun today as 'Rocket City' exploded in celebration.

Von Braun, clutching a mini-Stars & Stripes, was hoisted on to the shoulders of four local councilmen and waved to thousands of people gathered outside the Alabama town's courthouse.

Well-wishers held up banners with messages including 'Wernher, you're our boy!' and 'Huntsville is Rocket City'. To huge cheers, Von Braun told the crowd: "The spaceflights must continue. The ultimate destiny of man is no longer confined to Earth."

Doctor Wernher Von Braun is lifted shoulder high during the post-splashdown celebrations in Huntsville today.

There was a German flavour to the celebrations, with dozens of Von Braun's old World War II rocket team joining the throng before heading back to party in their mountainous Monte Sano community, nicknamed 'Sauerkraut Hill' by locals.

Doctor Bernhard Tessmann, originally from Berlin, said: "Ah! I finally feel relaxed now. Maybe we will celebrate a little now. Our friendships go back to the 1930s and we know each other better than we know our brothers and sisters. We'll come together like an old family. We've been very successful here."

NATION CELEBRATES

The Moon men were today invited to a special session of Congress to be held in their honour.

In Washington, the House approved a resolution commending the "magnificent team" of Apollo 11 and agreeing the astronauts should visit in August or September.

Throughout the nation, the scenes of excitement were similar to launch day, with many people crowding outside shop windows to watch on TV. In San Francisco, splashdown was marked by a fifty-gun salute at the city's historic Presidio fort and in Des Moines, state capital of Iowa, a reproduction of the Liberty Bell was rung for the first time since 1950.

Cheers erupted on the floors of the stock exchanges in New York. On a display which usually carries market prices, a message flashed up saying: "Astronauts Armstrong, Aldrin and Collins, so proudly we hail you."

In the town where Buzz Aldrin grew up – Montclair, New Jersey – the Clairidge Theater replaced the letters on its facia advertising the movie *Ben Hur* with the message 'Congratulations Buzz Aldrin, Montclair Man on the Moon'.

Montclair fireman Dan DiPrenda, who graduated from high school alongside Aldrin in 1947, said: "It's thrilling to think how I used to know Buzz and now he's coming back from the Moon!"

But in Salt Lake City, Utah, Mormons were marking the 122nd anniversary of their arrival there and seventy-four calls were received complaining when the local TV station interrupted coverage of a parade honouring Mormon pioneers to show splashdown.

USSR HAILS 'GLORIOUS DREAM'

The Soviet Union led the worldwide tributes to Apollo 11 today.

Moscow Radio took the unusual step of interrupting its scheduled programme to issue a news flash confirming splashdown. Listeners were told: "The courageous astronauts are again on our planet. We congratulate you on your return to Earth, courageous astronauts. Man's first flight to the Moon has been completed. The glorious dream of visionaries and scientists has come true."

Jubilant Mission Control staff wave miniature US flags today as they celebrate Apollo 11's successful splashdown.

For the first time in the mission, events were shown live on TV in the USSR and the nation's cosmonauts issued a statement saying: "With all our heart we congratulate you on the completion of the remarkable trip to the Moon."

The USA received dozens of congratulatory messages from political and religious leaders around the world. In England, the Lutine Bell at Lloyd's of London – recovered from a French frigate which sank in 1793 – was rung to mark splashdown. Traditionally it is rung to record momentous events at sea. Officials in Amiens, France, where *From The Earth To The Moon* author Jules Verne lived, said the astronauts are to be made honorary citizens.

Today's TV audience included US First Lady Pat Nixon, who watched in Honolulu because Navy protocol excludes women from its vessels during major operations. As the astronauts boarded the USS *Hornet* she exclaimed: "Oh God! They can still walk and all after being cooped up! Boy, it must be all right. We can tell by Dick's expression." Meanwhile, the widow of John F. Kennedy, Jackie Onasis, watched on a portable TV on board husband Aristotle's yacht in the Ionian Sea.

JAN: I'M STILL NUMB

Jan Armstrong admitted she was struggling to take it all in after watching the mission finale.

Jan, who knelt and prayed in front of the TV during splashdown, said: "If anyone were to ask me how I could describe this flight, I can only say it was out of this world. I'm still numb. I'm terribly, terribly proud of them."

Flanked by sons Ricky and Mark at a press conference outside their home, she added: "We look forward to being together, but the quarantine period is an absolute necessity. I have some plans for our first meal when Neil comes home. I know he's looking forward to apple dumplings."

HEAVEN HELPED HER

Joan Aldrin has promised to give God a break now the mission is over.

She watched splashdown at home with eight other astronauts' wives and, as the cameras scanned the sky for a sight of Columbia, she shouted: "Go find him, cameraman! Go find him!"

Joan told reporters: "I talked to my minister today and asked him if he'd been saying a lot of little prayers. He said 'Just one big one'. I said that's what I'd been doing, so maybe God can take a rest now!"

SPACE GROWTH

Pat Collins has given the thumbs-up to her husband's new look after he returned to Earth sporting a moustache.

She was as surprised as anyone when she got her first look at him on TV as he chatted to Richard Nixon, but she told reporters: "It's all right with me. He can keep it if he wants to."

She said: "I'm a very proud and happy woman. The whole thing was marvellous. I felt the greatest thrill at the first sighting of the craft." Pat admitted her children don't understand why their dad won't be back for nearly three weeks. "It's hard to keep children waiting... and wives, too!"

PROUDER THAN PROUD

Neil Armstrong's father was lost for the right words today as he paid tribute to his son.

Stephen Armstrong said: "I don't have words to describe how proud I am. It's not 'proud'. I'd have to get a dictionary to find a better word."

In emotional scenes outside their home in Wapakoneta, Ohio, Neil's mother, Viola, cried and said: "It's a wonderful day. I'm so relieved, so thankful. It's truly great." Asked if she'd like her son to make another Moon flight, she said: "I should say not!"

DAD SLAMS 'CRABS'

Space protestors got a rocket from Buzz Aldrin's father today.

Edwin Aldrin Snr told reporters Richard Nixon had "spoken for all America, except a bunch of crabs" when he praised the astronauts.

Speaking outside his home in New Jersey, he also took a swipe at the quarantine procedures, saying they were the idea of "some medicos" and "long-haired guys". He said: "Sometime in the future they're going to laugh at some of the stuff we do. Why in hell didn't they make the President wear a mask just to complete the deal?"

NEIL'S LOST IN MUSIC

Neil Armstrong proved how relaxed he is after returning to Earth…by strumming a ukulele.

His wife gave NASA the instrument to put in the quarantine quarters and photographer Lee Jones snapped him through one of the trailer windows playing it. Jones said: "I couldn't hear what he was playing, but he was grinning and looked like he was having a whale of a time."

The astronauts will spend three days in the Mobile Quarantine Facility (MQF) as it is transported by sea, air and road to Houston, where they will then remain in the Lunar Receiving Laboratory until 11 August.

The MQF has a shower, bunk beds, a kitchen and dining area. After living on space food, they'll enjoy hot meals thanks to another hi-tech device; a microwave oven. This cooks and heats food by exposing it to electromagnetic radiation. Early models, on sale since the mid-1950s, are too big for household use, but it's thought a new table-top version could become a popular kitchen appliance.

COLLINS PROMOTED

Michael Collins was handed a promotion today in recognition of his Moon exploits.

General John McConnell, Air Force Chief of Staff, sent a message saying the mission was "indeed a momentous achievement" and that Collins was being granted full Colonel rank as "a token of appreciation for the part you played". Collins was previously a Lieutenant Colonel and now holds the same rank as Buzz Aldrin.

SEARCH FOR OLD PROBE

The next men on the Moon will attempt to find an unmanned probe which landed there two years ago.

Within minutes of splashdown, NASA confirmed the launch date for Apollo 12 as 14 November, with a landing planned near the Landsberg crater, 800 miles to the west of the Sea of Tranquility. The site is close to the Ocean of Storms, where the unmanned Surveyor 3 soft-landed in April 1967.

Commander Charles Conrad and Lunar Module Pilot Alan Bean, both Navy officers, will make two Moonwalks and Apollo Programme Director, Lieutenant General Sam Phillips, said: "It may be possible to put down near enough to the Surveyor for the men to walk to it and examine it. It would be interesting to see what it looks like after two and a half years' exposure to the Moon's environment."

SICK OF SPACE FOOD

A Japanese TV presenter who spent the mission living like an astronaut admitted today he's relieved the experience is over.

Kokontei Shimba, 34, ate only three space meals a day and presented his show in a spacesuit. He complained: "The food was so tasteless I didn't know what I was eating. Twenty minutes after each meal, I was hungry again. I thought of the astronauts in space and of my wife's cooking every time I ate this tasteless food."

His bosses forked out thousands of dollars to buy the food from NASA and employed two bodyguards to ensure he didn't secretly eat anything else. He moaned: "The astronauts were confined to their space capsule, but I had to go to work every day and pass restaurants advertising delicious foods."

'NATIONALISM IS OVER'

A former Nazi rocket scientist who mentored Doctor Wernher Von Braun believes the Moon shot can usher in a new era of world peace.

Space visionary Professor Hermann Oberth outlined in his 1923 book, *The Rocket to Interplanetary Space,* how man could fly to the Moon. Yet, like his former student Von Braun, he found himself turning his expertise to making weapons of war for Adolf Hitler.

Oberth, 75, said today: "My dream is that the world should mature politically. Mankind has steered itself into a rather difficult position because his political and moral development hasn't kept up with his technology. It would be better now if there were co-operation between all nations in space exploration. The age of nationalism is gone. If there were togetherness in space exploration, there would be peace. People would see one vision and that vision wouldn't be war."

Oberth, from Nürnberg, West Germany, believes the astronauts deserve to be held in the same esteem as history's greatest explorers. "I know that other people say they're only a part of the huge NASA machine and are unlike Columbus, Lindberg and others, who did what they did alone. But, in my opinion, of course they are heroes."

ASTROS 'MUST VISIT GHETTOS'

A prominent civil rights campaigner says the Moon men should make visiting US ghettos more of a priority than travelling to foreign cities.

Writing in this week's African-American newspaper, *The New York Amsterdam News*, honorary Mayor of Harlem, Doctor Benjamin Watkins, said: "The world doesn't stop, even if a trio of astronauts get off it. Whether black people showed any interest in the landing is irrelevant. We in Harlem are demanding that the trio of astronauts include in their itinerary Harlem, Watts or some other ghetto, rather than Moscow or England."

The newspaper also carried an editorial comment about the mission, concluding with the line: "Yesterday the Moon. Tomorrow, maybe us."

GRANNIES IN SPACE

Airline TWA today formally applied for the US licence for commercial Moon flights.

TWA got a jump on its rivals by filing an application with the Civil Aeronautics Board to fly "persons, property and mail between points within the United States and points on the Moon".

The news came as the boss of NASA's astronaut team said the mission had defied those who believed a man couldn't survive the rigours of space travel. Deke Slayton, Director of Flight Crew Operations, said: "We always knew we could do it. At the start of the space programme we had to go through all sorts of medical and psychological tests to see if we could stand up to space; zero G and so on. Most of it was a complete waste of time!

"Anybody could go into space right now, just like they get into aeroplanes and fly. If we had a spacecraft big enough for passengers, you could just strap your granny in and wave goodbye, like you do at an airport. You could do it tomorrow."

D-DAY FOR MANKIND

Mankind has been urged not to turn its back on "destiny".

After blazing a trail to the Moon, space pioneers are ready to forge on to the planets. Looking ahead to future challenges, Doctor Werner Von Braun predicted today that the USA will commit within a year to go to Mars and insisted: "The Moon is just a commuter flight."

And Doctor Thomas Paine, NASA Administrator, warned that pulling the plug on the space programme "would be the end of the United States as we know it". He said: "This really is America at its best."

George Mueller, Associate Administrator for Manned Space Flight, made an impassioned plea for the nation to fund a manned landing on Mars in the 1980s. He said: "We now stand at what is undoubtedly the greatest decision of this planet. Four billion years ago, the Earth was formed. Four hundred million years ago, life moved to the land. Four million years ago, man appeared on Earth. One hundred years ago, the technological revolution that led to this day began.

"All of these events were important. Yet, in none of them did man make a conscious decision to follow a path that would change the future of all mankind. We have that opportunity and that challenge. For today, we conclusively proved that man is no longer bound to the limits of the planet on which, for so long, he has lived. We'll return to the Moon at regular intervals in the coming years, but these trips are only the first steps.

"There remains for mankind the task of deciding the next step. Should we substitute temporary material welfare for spiritual adventure and long-term accomplishment, then man will fall back from his destiny. The mighty surge of his achievement will be lost in the confines of this planet. This is the time for a decision."

CHAPTER 25

BACK DOWN TO EARTH
25 July 1969 and Beyond

Joan Aldrin speaks to husband Buzz via an intercom after the Mobile Quarantine Facility arrived in Houston.

BACK DOWN TO EARTH

HOMECOMING 10,000 people greeted the USS *Hornet* when it sailed into Pearl Harbor and the astronauts' wives were among 6,000 well-wishers gathered at Ellington Air Force Base, Houston, when a transport plane carrying the Mobile Quarantine Facility landed there. The astronauts were transported by road to the Manned Spacecraft Center to continue their quarantine.

FIRST WORDS Neil Armstrong cleared up the confusion over his first words on the Moon when, reading the mission transcript during his debriefing, he insisted there was a mistake in it and that he'd actually said: "That's one small step for *a* man …"

QUARANTINE No signs of 'Moon bugs' were found and the astronauts were released from the Lunar Receiving Laboratory late on 10 August, a few hours earlier than scheduled. Those quarantined with them were also allowed home, including technicians Ronald Buffum, 27, and George Williams, 42, who were placed in quarantine after being accidentally exposed to lunar samples. As the astronauts left, Buzz Aldrin told a crowd of well-wishers: "I need a little sun and a haircut."

CELEBRATIONS Millions turned out in New York, Chicago and Los Angeles to salute the astronauts during a twenty-four-hour coast-to-coast tour of the USA on 13 August. Richard Nixon announced that night that they were to be awarded the Medal of Freedom, the USA's highest civilian honour. On 16 August, more than 250,000 people attended a parade in Houston. In September, a parade in Wapakoneta attracted 50,000 people. In a world tour, the astronauts visited twenty-four countries in forty-five days.

LUNAR REFLECTOR On 1 August, scientists in California finally succeeded in hitting the retro-reflector left on the lunar surface with laser beams, accurately measuring the distance between the Earth and the Moon to within 150ft.

LUNAR SAMPLES Scientists found no evidence of living organisms and concluded that, although the Moon had been molten in the distant past, it was no longer volcanically active.

APOLLO 11: THE MOON LANDING IN REAL TIME

APOLLO FLIGHTS Public interest waned and only six more lunar flights were made. Five other landings were made at a range of sites, with three missions using the Lunar Roving Vehicle. Apollo 13 returned without landing after an explosion damaged the spacecraft. In all, a dozen men walked on the Moon. The last was Apollo 17 Commander Gene Cernan.

MARS Within a fortnight of Apollo 11's return, the world saw close-up photographs of Mars taken by two unmanned Mariner probes sent by NASA. They showed an inhospitable planet, with California Institute of Technology scientist, Doctor Robert Sharp, remarking of the rugged landscape: "Mars is a gal who wears no cosmetics." The USA chose not to press ahead with a manned mission.

SPACE SHUTTLE Although talk of a lunar shuttle, a lunar space station or Moon colonies never became a reality, NASA did develop a partially-reusable Earth-orbit shuttle. More than 100 flights were made between 1981 until 2011, when the programme ceased. Its success was overshadowed by two disasters – one shortly after launch in 1986 and another on re-entry in 2003 – which killed fourteen astronauts.

SOVIET SPACE PROGRAMME The Soviets never landed men on the Moon. They did continue their manned programme, however, and had a first joint mission with the USA in 1975, an Apollo and Soyuz spacecraft docking in Earth orbit. The USSR was the first nation to place a space station in Earth orbit; Salyut 1 in 1971. In 1998, Russia and the USA joined Japan, Europe and Canada to begin assembling the International Space Station.

RICHARD NIXON His approval rating soared by eight per cent in the week after the mission and he was re-elected in 1972. He resigned in 1974 over the Watergate scandal.

CHARLIE DUKE The Apollo 11 Capcom was Lunar Module Pilot on Apollo 16 and the tenth – and youngest – man to walk on the Moon.

DEKE SLAYTON One of NASA's 'Original Seven' astronauts, he was grounded because of an irregular heartbeat. He was later cleared to fly and, aged 51, was part of the 1975 Apollo-Soyuz mission.

ALAN SHEPARD America's first man in space was grounded two years after his 1961 flight because of an inner-ear problem. Reinstated to active duty in 1969, he walked on the Moon as Apollo 14 Commander in 1971.

JOHN GLENN The first American to orbit the Earth was elected to the Senate in 1974, serving for twenty-four years. In 1998, he became the oldest man in space, flying on the shuttle when he was 77.

MICHAEL COLLINS He retired from active astronaut duty and went on to pursue a career as a bureaucrat, including becoming Director of the National Air and Space Museum in 1978. He left in 1985 to go into business. He and Pat were together until her death in 2014.

BUZZ ALDRIN He left NASA in 1971, taking up a managerial role with the Air Force. He battled depression and alcoholism. Twice married after splitting from Joan in 1974, he became a prolific author. Remaining at the forefront of efforts to promote space exploration, in 2002 he punched a lunar landing conspiracy theorist who accused him of lying about landing and walking on the Moon.

NEIL ARMSTRONG He left NASA in 1971 and taught aerospace engineering at the University of Cincinnati until 1979 before gradually withdrawing from public life. He and Jan divorced in 1994. Armstrong married Carol Knight in 1999. The astronaut passed away in 2012, the first of the Apollo 11 crew to die.

BIBLIOGRAPHY

NASA archives; Life magazine; The Age (Melbourne, Australia); Altus-Times Democrat (Oklahoma); Baltimore Afro-American; Beaver County Times (Pennsylvania); Bonham Daily Favorite (Texas); Bryan Times (Ohio); Bulletin (Oregon); Calgary Herald (Canada); Chicago Tribune; The Day (Connecticut); Daytona Beach Morning Journal (Florida); Ellensburg Daily Record (Washington); Eugene Register-Guard (Oregon); Evening Independent (Florida); Evening News (New York State); Evening Times (Glasgow, Scotland); Daily Express (UK); Fort Scott Tribune (Ohio); Free Lance-Star (Virginia); Gadsden Times (Alabama); Gettysburg Times (Pennsylvania); Guardian (UK); Harlan Daily Enterprise (Kentucky); Herald-Journal (South Carolina); Lawrence Journal-World (Ohio); Leader-Post (Saskatchewan, Canada); Lewiston Evening Journal (New England); Lodi News-Sentinel (California); Daily Mail (UK); Meriden Journal (Connecticut); Michigan Daily (Michigan); Mid-Cities Daily News (Texas); Middlesboro Daily News (Kentucky); Daily Mirror (UK); Montreal Gazette (Canada); Morning Record (Connecticut); Mount Airy News (North Carolina); New York Times; Norwalk Hour (Connecticut); Observer (UK); Ocala Star-Banner (Florida); Owosso Argus-Press (Michigan); Phoenix (Saskatchewan); Pittsburgh Press (Pennsylvania); Press-Courier (California); Quebec Chronicle-Telegraph (Canada); Reading Eagle (Pennsylvania); Rochester Sentinel (Indiana); Rome News-Tribune (Georgia); Schenectady Gazette (New York State); Spokane Daily Chronicle (Washington State); Spokesman Review (Washington State); St Joseph Gazette (Missouri); St Joseph News-Press (Missouri); St Petersburg Times (Florida); Sumter Daily Item (South Carolina); Sydney Morning Herald (Australia); Daily Telegraph (UK); Sunday Telegraph (UK); Telegraph-Herald (Illinois); Times (UK); Sunday Times (UK); Times Daily (Alabama); Times-News (North Carolina); Toledo Blade (Ohio); Victoria Advocate (Texas); Virgin Islands Daily News (US Virgin Islands); Warsaw Times-Union

(Indiana); Washington Afro-American (Washington State); Washington Post; Waycross Journal-Herald (Georgia); Williamson Daily News (West Virginia); Windsor Star (Ontario, Canada); Youngstown Vindicator (Ohio).

ACKNOWLEDGMENTS

Thanks to my family – Vicky, Ella and Alex – for their encouragement and support and, along with Magnus Miles, giving up their time to read various draft versions of this book. Particular thanks to Vicky for her help with research and for revising the book as it came together. I am also grateful to Pen & Sword – and, in particular, Claire Hopkins – for their faith in the project and making it happen. Thanks to Janet Brookes for her work on the production of the book and to Jon Wilkinson for his excellent jacket design.

All images courtesy of NASA.

The Apollo 11 crew (from left) Buzz Aldrin, Michael Collins and Neil Armstrong with US President Barack Obama (far right) in the Oval Office at the White House in 2009 on the fortieth anniversary of the lunar landing. *Picture NASA/Bill Ingalls*